ARE ANGELS OK?

ARE ANGELS OK?

THE PARALLEL UNIVERSES OF NEW ZEALAND WRITERS AND SCIENTISTS

EDITED BY
PAUL CALLAGHAN
AND
BILL MANHIRE

VICTORIA UNIVERSITY PRESS

VICTORIA UNIVERSITY PRESS
Victoria University of Wellington
PO Box 600 Wellington
vuw.ac.nz/vup

National Library of New Zealand Cataloguing-in-Publication Data

Are angels OK? : the parallel universes of New Zealand writers
and scientists / edited by Paul Callaghan and Bill Manhire.
Includes bibliographical references.
ISBN-13: 978-0-86473-514-0
ISBN-10: 0-86473-514-6
1. New Zealand literature—21st century. I. Callaghan, Paul.
II. Manhire, Bill, 1946-
NZ820.8003—dc 22

Are Angels OK? received major funding support from the Smash Palace Fund, an
arts-science collaboration fund supported by the Ministry of Research Science
and Technology and Creative New Zealand.

Printed by PrintLink, Wellington

CONTENTS

ACKNOWLEDGEMENTS

We particularly wish to thank and acknowledge Glenda Lewis. Hers was the eureka moment which initiated the *Are Angels OK?* project. She also possessed the drive and skills to make it happen. We are grateful to the Royal Society of New Zealand for management support, and to the MacDiarmid Institute for Advanced Materials and Nanotechnology and the International Institute of Modern Letters. Thanks, too, to Tricia Walbridge and the Victoria University Foundation. Many thanks to Fergus Barrowman, Craig Gamble, Heather McKenzie, Sue Brown and VUP; to Rachel Barrowman and Jane Parkin for their scrupulous work as copy editors; and to Dylan Horrocks for his cover design. Kim Hill and Phil Smith of Radio New Zealand have been generous in their support, as have the staff of Te Papa, where a number of related presentations and performances were presented in November 2005.

Thanks also for advice and assistance to Faith Atkins, Matthew Parry, Toby Manhire, and Vanessa Manhire.

The Smash Palace Fund, which has made the production of this book possible, is a partnership between Creative New Zealand and the Ministry of Research, Science and Technology. Its aim is to encourage and support the convergence between the arts and science as a building block for innovation and creativity.

We are grateful to John McDavitt at Creative New Zealand for advice and support, and to our two referees: Martin Lord Rees, President of the Royal Society of London, and Sian Ede of the Gulbenkian Foundation.

Paul Callaghan
Bill Manhire

COPYRIGHT PERMISSIONS

INTRODUCTION:
BAGPIPE MUSIC

BILL MANHIRE

In her autobiography, Janet Frame tells of travelling back to New Zealand by ship and being befriended by a physicist – 'a mild, shy, pale young man' – to whom she gives, somewhat mischievously, the name Albert. Albert is travelling to take up a university post, and he brings her news of 'the world of science':

> I remembered how in my university days when I pored over the university calendars reading the science curriculum, I felt the excitement and mystery of the subject, 'Heat, Light and Sound', and I thought, 'Surely this is the province of poets, painters, composers of music, as well as of scientists . . . ?' And I remembered how, lured by the mystery and the magic, I borrowed books on physics and, opening them, I was faced with a wall of figures and symbols far beyond my Scholarship mathematics and chemistry: sentences that I could not understand, that roused the same feeling of frustration I'd had when I tried to read my father's bagpipe music . . . How could Heat, Light, Sound belonging to everyone, so remove itself from us?

Excitement and magic and mystery. Writers and physicists certainly have all that in common.[1] They are curious people who value the imagination. Both practise thought experiments, and thrive in the wonderful world of 'What If'. What if the earth went round the

---

1 'The most beautiful experience we can have is the mysterious. It is the fundamental emotion which stands at the cradle of true art and true science.' (Einstein, in 1931)

sun? What if we could travel back in time? What if an angel were suddenly to appear in a vineyard in 19th-century France? What if time and space were curved? What if we could look inside the atom? What if matter isn't solid? What if we put some writers and physicists together just to see what happens?

There are other similarities. One is a shared interest in the resonant power of words. It's not just quarks and black holes and starquakes that the world of physics has brought us (quarks, of course, were acquired courtesy of James Joyce). Physicists continue to produce phrasings as full of rhythmic energy as they are of apparent mass. Listen to how they talk: 'baryonic density', 'scattering amplitudes', 'temporal surprise', 'energetic outbursts', 'cloud chambers/bubble chambers/solid state devices', 'exotic particles', 'dark energy', 'event horizon'. The Nobel Prize-winning chemist and poet Roald Hoffmann has reminded us that the language of science is a language under stress. 'Words are being made to describe things that seem indescribable in words – equations, chemical structures and so forth. Words don't, cannot mean all that they stand for, yet they are all we have to describe experience. By being a natural language under tension, the language of science is inherently poetic.'[2]

The two groups share a further belief in inspiration, in what is sometimes called the eureka moment. Such moments equally come when the writer or scientist is working hard, or working not at all. Intuition is also part of what goes on. Sometimes this is simply a general alertness to the rewards of coincidence. Arno Allan Penzias and Robert Woodrow Wilson accidentally discovered background radiation, received a Nobel Prize, and so turned cosmology into a science. W.H. Auden's 'And the poets have names for the sea' was changed by the typesetter to 'And the ports have names for the sea'. Auden had the good sense to keep the error.

There is also a shared awareness of paradox and contradiction, and a sympathy for what Coleridge called the willing suspension of disbelief. Life inside the atom can be like certain agitated states inside the writer's head. The novelist Philip Pullman has pointed to a link 'between the way a writer can hold a repertoire of words, images, metaphors, in suspension before final commitment on the

2 http://www.pantaneto.co.uk/issue6/Hoffmann.htm.

page, and the mysteries of quantum physics, in which phenomena appear to exist in a state of unrealised potency until the researcher brings them into actuality by a specific measurement'.[3]

In the world of physics, aesthetic dimensions matter, too. 'A beautiful idea,' says Roger Penrose, 'has a much greater chance of being a correct idea than an ugly one.' Einstein's theories of special and general relativity are frequently praised for their beauty, for the ways they bring together simplicity and symmetry. 'Beauty is truth, truth beauty, – that is all / Ye know on earth, and all ye need to know,' said the poet Keats, going perhaps just a bit too far.

And then there is metaphor. It is hard to conceive of imaginative literature, poetry especially, without some metaphorical dimension. But metaphor in its broadest sense is a bridge between the unknown and the familiar. Physics needs metaphor and analogy to translate its understandings into terms the rest of us can understand. The Big Bang is not a directly articulated fact but a metaphorical explanation. Likewise a black hole. (Of course, some terminologies have the ability to outlast advances in scientific knowledge. Even a physicist can get up just after a pre-Copernican sunrise, work hard all day, and then sit on the balcony with a glass of wine, admiring an ethereal sunset.)

So there are differences and gaps. The biggest is there in the very language from which the physicist translates. It happens to be the language of a place where most of us cannot go. In this place, people converse in mathematics; there they can all play bagpipe music. Some physicists even think that mathematics is a language which pre-exists in nature, is indeed the language in which nature expresses itself. Mathematics is something we *discover*, not – like English or Russian or Cantonese – a sign system we devise to make our lives work better.

The other difference between writer and scientist can be found in the phase which succeeds the 'what if' question. The writer inscribes his or her imagined world, and tries – usually – to make it internally consistent. By contrast, as Paul Callaghan makes clear in his closing

---

3 'He refers frequently to the "Schrödinger's cat" conundrum, and what physicists call "the collapse of the wave function", in which a cat in a sealed box can be understood to be both alive and dead under certain abstruse quantum physical conditions.' Profile of Pullman in the *Sunday Times Magazine,* 24 October 2004.

essay, the physicist looks for empirical proof, trying to determine if the 'what if' can be clearly demonstrated, if it will fit with agreed observations and understandings. I suppose it is just possible to argue that the craft involved in *making* a novel or poem or play is somehow akin to this 'objective' process, but it is a pretty long stretch. In the end a poet, as E.B. White once said, 'approaches lucid ground warily'. This would not be a useful modus operandi for science. Whenever scientists have their heads in the clouds, they take care to keep their feet on the ground. As Alex Comfort[4] has written in his poem about assembling a competent team for mountain climbing:

> The particle physicist and the mathematician
> I'll take. They're crazy, but they watch their feet.

The *Are Angels OK?* project sprang from the energies of what many called the Einstein year, the 2005 International Year of Physics. It is not the first time New Zealand or other writers have ventured into the field of physics,[5] but it may well be the first sustained collaboration. The idea came from Glenda Lewis at the Royal Society of New Zealand. I don't think the plan was ever to 'explain' physics. Bill Bryson and David Bodanis and Simon Singh do this better than we ever could. But we thought it would be very interesting to see what would happen if 10 of New Zealand's leading imaginative writers were introduced to their scientific opposite numbers. We were interested in the collaborations and collisions which might take place. I made a list of writers, looking especially for variety of voice and genre.[6] Paul Callaghan made a list of physicists. We put our lists together and sat back to see

4   Yes, *that* Alex Comfort. The poem is called 'We want no dead weights on this expedition' and is printed in Maurice Riordan and Jon Turney (eds), *A Quark for Mister Mark: 101 Poems about Science*, Faber, 2000.
5   Margaret Mahy has been here before, while New Zealand writers such as the poet Cilla McQueen and the dramatist Stuart Hoar join others like Michael Frayn, Philip Pullman, Tom Stoppard, Lavinia Greenlaw, Frederick Seidel and Andrew Crumey in imaginatively exploring the perspectives and understandings of science.
6   Readers will see that within the unusual range of fiction and poetry here, there is also a comic strip by Dylan Horrocks, while Jo Randerson's contribution began life as a theatrical lecture. The writers' own comments, which form an appendix to their individual contributions, suggest just how far beyond their usual comfort zones many of them have moved.

what would happen. This book is part of what happened; we feel confident that much more will happen in the work these writers give us in the future.

Our title sounds thoughtful – romantic and offhand all at once – and Chris Price's fine verse essay gives it resonance and depth. Yet, in the best traditions of poetry and science, it is an accident. In an early, exploratory conversation, I expressed concern about how we might protect the writers' imaginative rights, and – based on my own uncertain brain – worried aloud about the dangers of woolly thinking. 'What if one or two of the writers head into territory that horrifies the physicists?' I said. 'What if they work within the sorts of understandings that quantum mechanics and the new cosmology have displaced? For example, are angels OK?' (It may have crossed my mind that angels are often messengers, figures of wisdom and rescue and redemption. But I doubt it.)

No worries, said the world of science: *Angels are absolutely OK*.

And so our project got under way.

Here is a story I once heard from a Dutch writer.

One evening two frogs fell into a vat of milk. One was a professor of physics. The other was a poet.

The professor of physics trod water for a while, then did a rapid calculation involving the buoyancy of his frog-body in milk. It was clear that he could not last. He gave a sigh and sank to the bottom, where he drowned.

The poet tried to remember what he knew about milk. 'Something something the milk of paradise,' came to mind. There was something, too, about the milk of human kindness. Some lines for a new poem of his own also occurred to him, though we will not quote them here. And all the while he went on treading water – or, more accurately, milk – occasionally wondering how long he could last.

In the morning, the farmer's wife came into the dairy. There in the vat was a large block of butter and . . . lying on top . . . a small, exhausted frog.

It would be nice to think this were true, but very probably both frogs died – or the physics professor found his way out through a wormhole.

Or perhaps they both trod water, kept up an awkward conversation though the night, and smiled at one another in the morning.

The physicist Werner Heisenberg says in his memoirs that science is rooted in conversations. He was thinking of the way scientists work with one another. I hope that this book can be one of the ways in which science has a conversation with all the rest of us.

Growing Space

Of course yours wasn't the first universe I've made.

God tinkers around in a shed at the back of the garden,
A retired artisan with a collection of vintage gear:
Coppery tanks, pipes, the smell of diesel and old kerosene,
An array of apparatus, disused tools and cylinders.

A bubble rises from froth and expands
And bursts, popping our ears. The machinery relaxes.

Ah well, that's what happens most of the time.
You've got to hold your tongue just so, not grip the tools so tight,
Even so, there's no guarantees in this business –
Pure chance anything turns out at all.

Some more adjustments, the foam pushes out another globule
Which inflates, fills with incandescence, floats out through the skylight
Suddenly bigger than the shed and inky blue around the edge.
The centre blazes brighter than a thousand suns.

This one could turn out interesting.

Tony Signal

UNOBTAINIUM

Elizabeth Knox

Naming rights

This story is more about family than time travel so I will start with my very particular grandmother. Grandma understood that her children had a duty towards her. None of them could ever manage to do the right thing by mere warmth or informality. There were things they just had to do, and they had to work out for themselves what those things were. For everyone in Grandma's life – family especially – there was an obstacle course of polite homage beyond which she waited with open arms. My mother was always reminding my father: '*You* have to call *her*. If she calls it won't matter how pleased you sound to hear her voice.' It was impossible for people not to feel that they'd neglected Grandma. The general impression she gave was of a woman who had been let down and had lowered her expectations. And still, on and off throughout the last quarter of her life, she let it be known that there were conditions she set which must be met.

In this story the one condition that counts is that a first son must be named Andrew. In the three preceding generations of Grandma's husband's family there had always been an Andrew, a first son of the generation. Because it wasn't a tradition of her own family Grandma seemed able to defend it with righteous inflexibility. Granddad had died young, when my father was only 15. Granddad Andrew, this was, and my uncle Andrew had read at his funeral. My father hadn't any investment in going against his mother's wishes; his name wasn't Andrew, so there was no danger of his first-born being an Andrew jnr. He and my mother rather liked the name and, when they were expecting their first child, the first of the generation, they agreed that, if it was a boy, they would call him Andrew.

My name is Andrew, Andrew McAlistair, but I have an older brother, whose name is Mark.

Lying on the lecture circuit

Two months ago I was at a conference with my colleague Tamara Glenbrook. Tamara and I are a double act: we deliver a two-part lecture.

I go first. Mine is on The Deity, and is adapted to the public facts, the palaver, the mystery and miracle of its existence. What it might mean that The Deity exists. How we had found it, how long it had been there, where 'there' is. What we mean when we talk about 'there'.

I don't want to collapse my story into the terms of my talk, but 'there' is tricky because The Deity is a wormhole.

I have an illustration on my PowerPoint of what we once imagined a wormhole might do to visible energy – the aura of its appetite. Everything in space is beautiful. An audience of lay people will sigh at a photograph of the horizon of the sun, or of the Horse Head Nebula. They sigh at the idea of The Deity because of what they can't see, its wound, the small vortex that creates a ripple in the gravitational field of the solar system – a veil over the Ark of the Covenant. On my PowerPoint I have diagrams of the kind of wormhole we think The Deity might be: a folded flat sheet of space-time, with a two-ended funnel milling through both surfaces, connecting distant places and times.

To tell the truth I never couple the words 'place' and 'time' in my lecture. I say 'place and period', because it is our firm belief that The Deity provides the possibility of time travel on a human scale. We do think we know when it might first have appeared – its downstream end. We have only to prove it. So – to tell the truth, in my lecture that is what I do; I tell as much of the truth as I can by saying 'place and period'. I allow myself that.

My lecture explains wormholes. When I deliver it I hear myself talking with less enthusiasm than I did when the real problem with wormholes was that there weren't any. Back then I'd be sharing the play of hypothesis, the vital intellectual project of testing our thoughts about the known laws of physics under imagined extreme conditions. (Occasionally back then I'd have a queasy moment of memory, of my high school, and being asked to explain – *smartarse* – the difference between 'hypothetical' and 'imaginary'.) The Deity is a real wormhole – mysteriously there – and yet still in my imagination. A *god* in my imagination because any portal to the past perhaps can do what God is said to do – save us.

My audience always wants to know how long it will be until we know positively *when* and *where* The Deity leads to. I always say, 'We are preparing experiments to determine just that.' And then I go on to describe the experiments, talking into their rapt silence. I describe the state of our knowledge six years ago. I tell them what we thought we might do back then. What I don't say to the gallery of softly-lit faces stacked above me in the auditorium, the choir whose silent attention is a song of wonder, is that we've *already* performed our first experiment and are only waiting to hear back.

I lie to them, because secrecy is essential at this point in the project.

Tamara then takes her turn. Her talk is about the ethics of time travel – though we aren't talking about even hypothetical travel in time, only about sending signals to the past.

Tamara's job is to prepare the ground, for if we ever do hear back, each one of these people is a potential petitioner. Everyone has something they want to fix. Everyone has a warning they'd like to give. Even altruists will want to send warnings about – say – climate change.

The gist of Tamara's talk is this: if we discover some method of sending signals into The Deity, signals that might emerge, coherent, at the other end of the wormhole; if we send a message that will provide us with evidence, some sign of ourselves, our presence in the historical record, what are we doing? What might happen? Is what we are doing the metaphorical equivalent of this: building a door for which there is only one key, a key we then send back in time, then wait to find, to find in history, so that we can unlock our door? Hasn't the key then come to us before the door? Won't we then have a key for which no lock exists, so that we'll have to build a lock in order to try our key for size?

Tamara tells her audience that the universe of this kind of time travel is relentlessly consistent. It's a universe in which there are always things we will do, because we've already done them.

The other possibility is this – we *don't* hear back, the key doesn't arrive in the post of history, there is no sign of it in the archives.

Tamara and I know that one of these things has already happened; either our message never got there, or something in it so changed things then that the 'now' following on from then isn't *our* now. The old past has created a different present and, with it, a different universe. In this scenario we could shout anything down the wormhole's throat and it wouldn't make any difference to us, because all the differences would generate new timelines. The Deity would turn out to be something like a bubble pipe – whenever we breathed into it, it would blow a bubble of another universe.

So, Tamara tells the softly-lit, expectant faces, in effect either everything we do is what we were *fated* to do, or nothing we do makes any difference to us.

She says that, in some time travel stories, the time traveller makes a mess of his own history and then keeps trying to fix it. He remembers his mistakes. He does things over and over. He makes adjustments till he is sick with possibility. But these are only stories. Stories have points of view, reality does not. None of us should imagine that, as a result of our experiment, we will be landed with several sets of memories of how things went. This, in effect, is the Copernican Principle. Copernicus proved that earth wasn't the centre of the universe. Then, much later, Edwin Hubble showed us that the galaxies were distributed equally and expanding away from us at the same speed. The Copernican Principle tells us that the expansion of space is the same relative to everyone and everywhere. And so, if none of us is special, then none of us is a default setting – the one who will at least remember how things were before the changes commenced.

Grandma's deathbed

She squeezed my hand. I looked around and tried to catch my father's eye. He, my mother, and my older brother, Mark, were out in the hallway with the cardiologist. Mark was arguing with the doctor.

Mark at 22 was a more confident version of his present self. Back then I still imagined he might mellow someday. I was waiting for it to happen, as people wait for a flooding river to abate. Mark would ease off, I thought. It was only a matter of time, and dependent on the conditions of his life. Surely he'd be different once he finished his degree, got a real girlfriend and a proper job. I wasn't exactly giving him the benefit of the doubt, it was more a case of my still seeing him the way he saw himself – as an eccentric hero and the only real honest soul. It was my habit to see him that way; he'd been so persuasive, so all-pervasive in his views.

My brother had a theory about Grandma's treatment, which he'd aired to family only for the last hour at her bedside. His theory was this: those doctors weren't doing everything they could. He'd read about a new drug therapy. The treatment was time-sensitive, and those doctors would have to start it right away.

I could hear Mark in the hallway telling the cardiologist what he had been telling us. His voice was loud – not raised, for its default setting was loud. He was repeating to the doctor everything he had said to us – all of it, word for word – for précis was impossible for Mark. 'There is a new drug, isn't there?' he said. 'They could try that, couldn't they?'

He didn't say 'you' to the doctor, or look him in the eye.

'It's about the money, isn't it? Hospital policy, the overheads.' Mark's hands went up and flapped by his face, so that he looked like a minstrel from old music hall routines, gloved white by the fluorescents. I guessed that he thought he looked sarcastic. But, for a young man whose Grandma was dying, standing beside his father, whose mother was dying, he only looked mad.

The cardiologist was astonished, but began to reason with my brother. The cardiac ward was itself an extreme condition, and no doubt this doctor had seen people misbehave themselves at deathbeds before.

Mum was very distressed by Mark's carry-on, so distressed that her human instincts drove her, against 22 years of counterintuitive biofeedback, to place her hand on my brother's arm.

Mark flinched and snatched his arm back. He began to tremble, but he kept on talking. He had dredged up the name of the new drug. His voice cranked up a notch and in it, just detectable, was a hint of a boast about his recall, about what he knew, an eagerness completely out of keeping with the deathbed. His tone said, 'Here I am, not a doctor myself, educating the doctor, or at least letting him know that *I* know there's more he could do if budgetary concerns didn't prevent it.'

The doctor said, plainly, that the drug wasn't suitable in these kinds of cases.

Mark went on as though he hadn't heard.

A nurse came into the room and looked at the bag that was collecting what came from the catheter. She tapped the plastic tube, and the smallest frown possible flickered across her face. She didn't mean me to see it, but I saw it, and knew that Grandma's kidneys were packing up.

Then Grandma squeezed my hand again. I looked at her and found her eyes open. 'Andrew,' she whispered. 'Is that Mark?'

'Yes,' I said, and turned away again to call out to my parents and brother.

'No,' said Grandma, very quiet. Then, in a whisper, 'Andrew. I'm glad you're Andrew.'

Her eyes wandered.

Everyone came into the room, and everyone but Mark put a hand on her. Then Mark did, watching us all. He folded his hand over her ankle under the waffled cotton blanket. Grandma lowered her eyelids and stretched her throat.

'Excuse me,' said the doctor, and edged me aside.

The nurse and he stood for a moment on either side of the bed at Grandma's head. They had their hands raised as if they meant to catch something.

Nothing happened. Grandma settled again and the doctor and nurse stepped back and we stayed there and I watched the numbers on a monitor slowly, slowly fall.

Later Dad asked me what she'd said, and I didn't tell him. Or, rather, I made sense of what she'd said by telling him she said she was glad to see me. 'Oh, and she said, "Is that Mark?" because she could hear him.'

'Him too,' Dad said, and his eyes filled with tears.

'That's right,' I said, because Grandma was always terribly hard to please and it hadn't mattered to me because grandchildren don't have to do very much to please their grandparents. But Dad had had a difficult time with her – for many years she *was* his time, and the difficulty was like a deformed leg or bad lungs, some trouble he'd laboured under all his life. Grandma was difficult, and often ungracious, and by saying she'd said she was glad to see me I meant to claim that she'd blessed us, me and my brother. Dad would love her for doing that – blessing his sons. Also, I did imagine that was what she meant to do, bestow a blessing, *my* blessing: 'I'm glad you're here, Andrew.'

However, if I had reported faithfully her actual words, instead of translating them into sense, I might have had the story sooner – and sooner understood a salient fact about how the universe worked.

For whatever *that's* worth.

Buttonholed

Tamara and I concluded our lecture by answering questions. Then our audience filed out, smiling lovingly at us. Some stayed, hovering by the podium where we waited while a technician detached our laptops from the audiovisual equipment. One waiting bunch were the people taking us out to dinner, the others were those with questions they'd been too shy to ask at question time.

The lights came all the way up once our laptops were uncoupled. Beside me Tamara cocked her head to catch what someone was saying to her over the racket of the theatre's tilting seats twanging up empty.

A man shuffled towards me and only stopped when his belly brushed my computer bag. His sour smell enveloped me. He started to talk. I tried to meet his eyes, which seemed to have extra thick corneas, their colour and character fixed and suspended under a deep glaze. His jacket was grubby. I felt my eyes slip out of focus. It was what I had learned, defensively, to do.

He was saying, 'Have you ever thought . . .', nervous and eager. He rummaged in his shoulder bag and came up with a thick manila folder. He had a theory, he said, he just needed someone to do the maths.

I tried not to see the orange oxidised egg yolk on his sweater. I attempted to interrupt him. 'Listen,' I said, then, *'Listen!'*

He finally petered out, though his lips kept moving as if he were sucking on some invisible tit. He gave me his attention, peered at me through eyes like those of a shying horse.

'Do you think your theory can be up to much without the maths?' I said. And he immediately started up about Einstein and intuition and how he was more imaginative than intellectual.

He had receding gums, and a mortar of tartar between each tooth. I thought of Mark in high school explaining to me how the *average* teenager was hung up on externals, on all the conventions about how to dress – clothes, hair, make-up, deodorant, the whole superficial works – while never seeming to care whether they had anything interesting to say.

'Hang on,' I said. 'When Einstein was riding on a tram looking at the clock on the Bern town hall and wondering what would happen if the tram was going at the speed of light – he could at least do the maths. He had a long journey ahead of him, but he did have legs to walk on.'

The buttonholer got a sly, smug look. 'But didn't Einstein's teacher Minowski do a lot of it for him? Minowski's calculations are essential to the Special Theory of Relativity. And famously elegant.' He bobbed up and down, pleased to have trumped me, and to have summoned the facts about Einstein's teacher. Again he proffered his folder. 'Won't you at least listen to my . . .'

'No,' I said, abrupt. 'I've done enough listening for a lifetime.'

In the car on our way to the restaurant Tamara said to me, 'Andrew, I don't know that I've ever seen you be so short with one of those awkward enthusiasts.'

'He was a particularly odoriferous specimen.'

'Did you imagine you'd catch something?'

I thought of the pebbly surface of the floor by the sink in the kitchen of Mark's house, where rock hard potato peelings were adhered to the lino. 'I never imagine that.'

Tamara was a philosopher, a physicist and ethicist. She said to me, 'Are you saying that under no circumstances would you get squeamish about something or someone unsanitary?'

'Look, Tam, I'm not defending my prejudice. But if you knew my brother . . .'

'No one knows your brother.'

'That's because he's not in the same world as me.'

'Really? Is he on the far side of The Deity?'

'The receiving end? That's the last place Mark would ever be. Mark, in defiance of the Copernican Principle, has always been special. The universe expanded away from *him*.'

Tamara looked at me for a long while. Then she said, 'That message you sent yourself, it was about Mark, wasn't it?'

Perturbation

The Deity – or Mr Ed, as we of the team familiarly call it – is a natural phenomenon. It seems that, at the beginning of the universe, the Big Bang, matter and anti-gravitating exotic matter bundled themselves up, one encasing the other. Together they formed a small planet-sized body of zero mass. It may not be the only one out there, invisible to our instruments, but so far we've only found the one. We believe that, like the planets, The Deity has always been there, a feature of our solar system, in the outermost orbit, its gravity and anti-gravity so perfectly balanced that, up until seven years ago, no net effect was created on the gravitational fields of the system.

Seven years ago a long-period comet from the Oort Cloud visited our system. Astronomers all over the world closely observed Ranchod's Comet. Its trajectory offered no cause for alarm. Professionals and amateurs simply enjoyed it. It inspired no eschatological thoughts, no Heaven's Gate cult hoping to hitch a ride to Paradise on its tail. Ranchod's Comet passed around the sun at the beginning of the southern hemisphere spring, and picked up acceleration, slung away from the sun's gravity. It passed back through the system, closer to earth on its way out.

My family and I spent several evenings lying on our lawn and looking up at it – a bright quotation mark, its tail as straight as a gas plume.

Then it went beyond the reach of the naked eye and only astronomers peered after it.

And then it vanished, was snuffed, was swallowed whole.

In the days after its disappearance, telescopes, from hobbyists' Skywatchers and Celestrons to Hubble itself, peered at the region where the comet had been and wasn't any more. They couldn't see a thing. In fact the first people to see what was there instead of the comet were a group of scientists in a laboratory in the Cashmere Hills, near Christchurch, in my home country. The scientists had a ring laser – two counter-rotating laser beams fired in a circular path – an instrument that measured gravitational fields. It was so sensitive that it could measure earthquakes on the moon. What this ring laser detected was a perturbation in the gravitational field of the solar system. The perturbation wasn't caused by an unknown planet or moon – the spatial character of the gravity field was all wrong for that. The Cashmere Hills mob said to themselves, '*What is this ripple?*' They said, 'Isn't it coming more or less from the same place where Ranchod's Comet vanished?'

Then they picked up the phone and called NASA.

NASA had an array of ring lasers in satellites orbiting Earth – the 'seismic spies' of conspiracy theory web sites. The array was able to get directionality, able to nail the source of the ripple. NASA aimed their array at the dark region beyond the orbit of Pluto and found that the ripple emanated from a vortex in – well – in apparently nothing.

By now every cosmologist on the planet was sitting up, eyes wide open, hair standing on end. Out there was a perturbation in the gravitational field caused by the fluctuations of a vortex. The vortex had a position. Its position corresponded to the last known position of Ranchod's Comet. The vortex was where it might be were it a feature of an unknown astral body with which the comet had collided, collided and been swallowed. These were the facts.

Obviously – thought the cosmologists – if 'it' was there (and it was) and it had zero mass, it must formerly have been in a state of perfectly balanced gravity and anti-gravity. We knew that anti-gravity existed, at least at the quantum level. We knew that there were anti-gravitational quantum particles. But anti-gravitating matter in large quantities was always the 'unobtainium' of theories – theories about how the throat of *a wormhole* might be kept open.

Physicists discussing wormholes always faced the problem of how to actually make one, how to turn the ravening singularity of a black hole into a gravitational vortex with an open throat, so that whatever went down it wouldn't simply be torn apart or annihilated. One hypothesis was that anti-gravity would be required to keep the throat of the wormhole open and stable. Now here – we all supposed – was a stable zero-mass body. That suggested anti-gravitating exotic matter. The body had a hole in it. And something had gone into the hole and apparently had not emerged again. Had this formerly invisible astral body been the ungerminated seed of a wormhole, always there and invisible to us?

'Always there and invisible to us, like God,' some smartarse said. Then someone else called it The Deity, and the name stuck despite later squabbles about naming rights, and embarrassed attempts to translate its now popular name into some antique and supposedly more dignified language like Etruscan or Toltec.

The Deity had swallowed a comet, emitting no debris or radiation in the process. The comet had left a tiny wound in the shield of its balanced gravities. The hole remained open. Anything else that went into that hole – an aperture that we eventually calculated was only around two metres in circumference – would, presumably, vanish as thoroughly and finally as Ranchod's Comet.

Shame and scientific endeavour

It was a man called Eric Hall who conclusively decided the matter of what The Deity was, and what might be done with it. Hall wasn't a cosmologist. He was an historian with an undergraduate degree in physics, whose life's work was the history of late-modern astronomy. His period was World War Two to Glasnost. What Hall did was *find* Ranchod's Comet. It, too, was already there (but not always there) and in the historical record.

I want you to imagine a journey. Nights, days, then an afternoon.

Eric Hall drove from his home in Ohio, near Ohio State University and the old Big Ear radio telescope. He drove across country with a folder containing 20 pages of handwritten notes, plus photocopies of pages from the logbook of Mt Palomar Observatory, and correspondence between an astronomer at Mt Palomar and the National Geographic Society. Hall had possessed the photocopied material for years. It lived in a file box in his study. He had cited it in a footnote to an article he'd written some years before for the *American Scholar*, a piece called 'Shame and Scientific Endeavour'.

Imagine this man, an historian, virtually a scientific-minded layman, his hands gripping the steering wheel, his suit jacket on a hook in the side window, his notes on the passenger seat beside him.

He didn't send an e-mail. He didn't pick up the phone. He didn't even think to get on a plane. He drove across country as a way of keeping his feet on the ground. What he had to say was so serious that a crucial sense of secrecy, a secrecy approaching the sacred, compelled him to communicate what he suspected, not by phone or e-mail, but face-to-face. He hadn't even called ahead. No – wrapped in a kind of holy terror Hall drove from Ohio to Massachusetts to talk to the man who, it seemed to him, was making the most steady authoritative sense so far of the new phenomenon.

Hall was lucky. It was term time and the cosmologist he wanted to speak to was teaching and was there when he arrived.

Imagine a car park, black asphalt gritty with salt. This is Amherst, so, as we follow Hall into the Faculty of Physics, Emily Dickinson will walk alongside us.

There's a certain Slant of light,
Winter Afternoons –
That oppresses like the Heft
Of Cathedral Tunes –

Heavenly Hurt, it gives us –
We can find no scar,
But internal difference,
Where the Meanings, are –

Hall climbed the stairs. The landing gave him a view only of the very top of the old chapel. He was unable to make out whether the ironwork object on its spire was a crucifix or a weather vane. He stood frozen for several minutes, staring, because he felt he needed to know. He had become a superstitious man.

Eventually he got to the right floor, and the right door. He sought, and he found. He knocked, and it was opened unto him.

This is the discovery Eric Hall outlined in his notes, and illustrated with his photocopied supporting materials.

In April of 1956 an astronomer looking through the eyepiece of the Hale telescope at the Mt Palomar Observatory saw a comet appear. His reasonable assumption was that he'd only just noticed it. He showed it to the other astronomers with him that night. They called other people to log their sighting – hoping they'd get to name it. They watched it over several nights and did what people do with comets – they mapped its trajectory. Since they had only just noticed it they assumed that it was heading into the solar system. But, after several days, it became clear that it was not. Indeed, it was moving *out* of the solar system, on a trajectory that indicated it must already have passed through, made a circuit of the sun, and was now heading back into interstellar space. The Palomar astronomers wondered, how had they possibly not noticed this object earlier? *How the hell* had it got by them? Surely, when it passed near to Earth it must have been visible to the naked eye.

These astronomers hadn't ultimately made it into Hall's paper 'Shame and Scientific Endeavour', for they were more embarrassed than disgraced. It was the opinion of their colleagues elsewhere that they'd got their calculations wrong. Obviously. No great shame in that, or even terrible embarrassment. They should just fess up and say sorry. But instead they kept insisting they were right. The comet seemed to have simply appeared from nowhere. One of them even stuck to his guns for years, whenever the subject came up, to the point where he became a bit of a joke an astronomical circles. (Also, once when drunk he had notoriously raved on about a 'strange green light in the sky'.)

What Hall said to the cosmologist in Amherst was, 'I think the Mt Palomar boys' "comet from nowhere" is Ranchod's Comet. I think that after it went into that shy singularity out there – whose name I refuse to use – it came out in April 1956.'

'April 1956, *in this universe*,' the cosmologist said.

'Yes. In this universe. So you're right and it is a wormhole. And I suppose we might call it a short-period wormhole.'

'I suppose we might,' said the cosmologist – mightily pleased.

He called me that night. My daughter fetched me in from the barbecue. I hadn't seen Edward since a conference at Stanford more than a year before. 'Hey Ed!' I said, 'how are you?'

'Overexcited,' he said. Then, 'Andrew, for years you've been thinking and talking and writing about time travel.'

'Yes?'

'Well – now the time has come.'

The Massachusetts Project

We put a team together. It was quite a small team. From the start we knew that what we hoped to do was risky and would be best done with discretion. We kept it small, and kept our mouths shut. We thought we might have to drum up funding – vast sums of it – but quite quickly discovered that our experiments were relatively inexpensive. We did have to put in a lot of time, and needed funding for that. We had to work in close proximity to one another and *keep things in the building*, so to speak. Because Edward had a happy relationship with his university we began at Amherst. There were only five of us then – just thinking and sometimes consulting other interested parties, everyone who was gazing in wonder into that place beyond Pluto, at the broken symmetry of that mysterious mass.

Once we'd worked out what we wanted to do and how to do it we recruited a bunch of other close-mouthed colleagues, and asked the permission of those who know their permission is always required – the United States government. We moved to a laboratory in New Mexico – where the skies were reliably clear. We chose our digital modulation code, and built our Photon Telegraph. We devised our message protocols. We even began to *send* our messages.

Within two years of the discovery of The Deity the 'Massachusetts Project' was up and running. After that all we had left to do was wait, and to go on pretending to the public that what we had done was still only what we planned to do.

Time in narrative is a matter of sequence, frequency and duration. For a long time we frequently represented our actions in the wrong sequence. And that is why, nearly seven years after Edward's phone call, I'm still sometimes giving lectures about The Deity and the ethics of time travel. I am lying: laying out as optional the inevitable.

(Tamara Glenbrook, cosmologist and ethicist, sitting beside me in the back seat of a car pulling into the curb beside Nathan's in Georgetown, D.C., asking, 'How come you never talk about your brother?')

'Singularities are pathologies, of course, though not all pathologies are singularities'

Mark and I sit watching television. I'm very little and the seat of the chair I'm in is so long it doesn't let my knees bend. I'm sitting as straight-legged as a doll.

We are watching *Mr Ed*. It has just started and the theme song is on. *'A horse is a horse, of course, of course, and no one can talk to a horse, of course, that is of course unless the horse is the famous Mr Ed!'*

(I'm not sure what year this is. *Mr Ed* screened later in New Zealand. Everything screened later here. We've always got first light, but in the 60s everything else came in delayed broadcast. For instance, no one in the Department of Education was up with the very latest in abnormal developmental psychology.)

Mum and Dad let us watch *Mr Ed* because it was 'good for Mark'. Mark warmed more to stories about animals than those about people. He and I used to play at *Mr Ed*. I was always poor, befuddled, put-upon Wilbur. My Wilbur would sometimes get fed up and walk away from the stable door. Mark's Mr Ed would follow, the door no impediment to him. Once Mark was in full flight, in or out of character, he was unstoppable. He would monologue for hours at a stretch. But he was often an entertaining child, and his Mr Ed wasn't a wise ass, more a Jester, poking fun at human foolishness. At eight Mark was already creating his catalogue of silly things ordinary people do.

Mark at eight, at our Hawke's Bay primary school. While the other eight-year-old boys were building forts and kicking balls, Mark was walking up and down the row of Arbor Day plantings, waving a stick, and talking. He was talking to me. I was playing marbles, a big girl trying to teach me how to 'flick' and not 'funk'. I was listening to my brother; it was my habit. I never did learn to tune people out till I had a chattering (but normal) eight-year-old of my own.

Mark wasn't normal. My parents never supposed that 'normal' was something to aspire to, so they managed his peculiarities kindly and with as little panic as possible. Dad once mentioned to me that when Mark was three they bought him a full-length mirror. 'We hoped to catch his wandering gaze,' Dad said. 'Mark would scarcely ever meet his mother's eye. We hoped to see him meet his own.'

The school wasn't as easily troubled by Mark as our parents were – schools are blockheaded and thick-skinned. But it was far more determined to get to the bottom of Mark's awkwardness with other children, and his intransigent inability to listen.

This was the 60s, and this was New Zealand, and the child psychologist they brought in to see my brother didn't know what he was looking at. Mark had a wonderful time with the man; loved the tests, loved the talk. He was always happier talking to adults.

The psychologist decided that Mark was only a highly intelligent child, and obviously playing up because of a lack of stimulation. Mum and Dad's solution to this was to sign Mark up for all sorts of extracurricular activities. They found a speech and drama teacher, a chess club and art lessons. My father, a jazz enthusiast, conducted informal music-appreciation lessons. Our parents hadn't much money so they took turns escorting Mark to the plays and operas they might otherwise have been able to attend together.

I did benefit from some of this. I shared the paint boxes and good art paper. I was entertained. Mark acted out *The Mikado* for me. He taught me chess, and regaled me with recitations of whatever he learned at speech and drama – I remember Robert Louis Stevenson and Alfred Noyes.

Mark was happier. He was excused and defended. The teacher he got the following year was aware that she had been given a gifted child and when the other kids teased him she'd shout at them that what they had to understand was that Mark was a genius.

Excused, defended, and sufficiently explained to himself; set apart and *above* others.

Mark loping along a corridor at school deep in thought. His gait is uneven, a kind of cantering, and he's flapping his hands. He's excited and his eyes are bright but glassy.

By the time I reached the age when I might be embarrassed by my brother I had signed on to the 'different is special' belief system. I spurned kids who failed the test of Mark. I didn't often have to rise to his defence for he didn't notice wariness or discomfort in others, he only discovered that they didn't approve of him when they were openly sharp or censorious. There were times when I had only to stand back and watch, and perhaps even profit from, his callous enthusiasm. For instance, when he could not be made to see why our friend Jay couldn't come over since Jay's mother would have to come to collect him, crossing town in the rush-hour traffic. 'It just doesn't suit me,' said Jay's mum, falling back on her adult authority, because she could see that Mark would keep refusing to understand what she had so patiently explained to him. Mark kept pushing, he wasn't being insolent or impolite, he just kept asking and not listening to her answer. Jay came with us. Jay's mother called my mother who drove Jay home through the rush-hour traffic, because it was vitally important that Mark keep the few friends he had.

Mark didn't understand individuals but, as a child, he was imaginative and systematic in describing groups of people. He was a playground anthropologist. He could work out the whole plot of a group, or at least tell a good story about them.

There was the group who didn't like ball games and shared the hurdy-gurdy, a pole with a swivelling top to which a rope was fastened. We'd take turns running and swinging on it. But one day a ball smashed into my friend Selina when she was spinning, and broke her glasses, which cut her eyebrow. When we came to school the following day the hurdy-gurdy's rope was coiled and knotted around its top. It was out of bounds.

Mark's take on the situation was this: that although the hurdy-gurdy was fixed in its corner of the playground, while the ball games wandered untidily everywhere, the hurdy-gurdy was deemed the problem. It was a problem because the kids who liked it weren't 'fitting in'.

That was Mark on majority versus minority culture. He said, 'It's never the fault of the boys with the balls. What *they* think is this . . .', he would say, and lay out his theories about the thinking of teachers, or the way the school was run. He was astute. He often understood what was going on. But he never understood how people felt about what he said, or how, when he continued to say it, it didn't make it any more likely – and maybe made it *less* likely – that his case would be heard. He never seemed to understand this, despite being told directly. Our mother would say, 'Mark, if you won't stop talking no one will hear you.'

*

Mark turned on to physics because of stories about its brilliant heroes. I remember him coming home with a story about Einstein. He told it to Mum and me while Mum was mashing carrots and swedes in a pot and I was watching the steam pressed out as grainy clouds, like spores from a dry puffball.

Mark said, 'Did you know that Einstein didn't talk at all until he was three years old? The first thing he said was: "Mother, my head is cold. May I have my hat?" It was one sign of his being a genius. His mind was simply on other things – that's what I think. He might have waited even longer till he really had something to say, eh? Only his head happened to be cold.'

'And his mother's mind happened to be elsewhere,' Mum said, wry.

'So Einstein was like Mr Ed,' I said. Then sang, *'People yackity-yak a streak and waste your time of day, but Mr Ed will never speak unless he has something to say!'*

Mum said, 'You talked very early, Mark. It was Andrew who talked late. We were beginning to think he was very stupid.'

'But instead he was very normal!' Mark said, gleeful. 'Andrew and I are Einstein's theories of relativity. I come first. I'm Special and he's General.'

'Very witty, dear,' Mum said.

Physics became Mark's thing – the latest in a succession of obsessions. Mark's enthusiasms were a deep, prolonged communion. He owned his subjects, soaked them with his scent. He would say that he liked other people to be interested in what interested him, but that only meant they must listen to him.

Mark would tell you what he knew. He'd tell you what *you* knew. He'd tell you what he'd already told you. But he didn't like anyone to talk about his subject. Or, rather, he could attend a lecture, but friends and family weren't allowed to show that they'd learned from the compulsory association with whatever it was he lately loved. Mark's subjects were his world, floating islands whose fundamental aqueous instability he'd feel only whenever anyone else set foot on them.

*

By the time I was 12 I'd learned to steer clear of knowing much about my brother's subjects. He was always telling me the same stuff over and over again, so he must have been pretty confident of my ignorance. By 12 my battle fatigue at my brother's frequent 40-minute dissertations sometimes even caused me to shudder at the sight of his books. So I professed not be interested in physics, hoping to sidestep his lectures.

As Mark entered his teens his talking had become more intense, more toneless, less directed at whomever he was speaking to. When he was young, and genius pardoned his peculiarities, he didn't mind not fitting in. As a teenager things became more difficult. For instance, he confounded being fashionable with being clean, and refused to be fashionable. He went unwashed. My parents fought with him and he wept and kicked holes in our walls and demanded to know why he couldn't be loved just as he was, and then raged on about their 'conformity' and 'middle-class expectations'.

When I followed Mark to university I was doing information science, maths, and a stage one physics paper my course coordinator talked me into taking. Being a first-year I had to take the slim pickings of lab time on the computers. (This was back in the early 80s when computers weren't two a penny.) I spent longer each day on campus than I otherwise might have. I didn't think much of my physics paper so, in my idle hours waiting for the clock to come around, I attended some third-year lectures to see whether I should go on with it, or drop it.

By that time, the start of my second term, Mark had dropped out of university altogether. Though physics was his thing, it turned out he wasn't very good at maths. Suddenly he was required to make the 'consistent effort' his high school teachers used to talk about – to his scorn. Mark had had two-and-a-half patchy years of effort, but was too disorganised to get assignments in on time, and too combative to be rewarding or even reasonable to teach. He declared that 'formal education' wasn't for him. And it wasn't. For Mark going to school was never about what there was to know, but about dazzling people with what he already knew – dazzling not to win them over, but to defend himself against them.

*

(Oh Mark! – Mark at 10, feeling got-at by a teacher, coming home and saying to Mum and me, 'When I grow up I'm going to change my name to "Example", so that things will always be *for* me and never against me.')

As we grew older and lived apart and things continued to fail to work out for him, Mark's efforts to pull a younger brother's thoughts back into line with his were often insensitive to the point of brutality. He would shout at me, 'The way you see things, Andrew, is this!', then go on to represent someone I couldn't recognise as myself. If he did happen to hear and take in anything I said, he'd get this look on his face of outraged incredulity, and would say: 'I don't know how you can say that', then, later, 'Oh, well, *you* would say that.'

In the end it was just too lonely that I could never say what I thought. And I was exhausted by the years of hours and hours of talk. I hadn't been able to disengage myself. You see, this was my brother, the companion of my childhood. How could I choose not to listen to him and wait my turn? *When* would I have chosen that? When would I have known to do things differently? This was my older brother, the companion of my childhood – to whom I was invisible and inaudible.

Go right to the source and ask the horse

The New Mexico lab, one evening. My boss Edward came in when I was watching some technicians tinker with the Photon Telegraph. None of us heard him till he said, 'What are you whistling?'

We were whistling the theme song from *Mr Ed*. I'd started it, so confessed.

'Remind me,' Edward said.

'Go on, Dr McAllister,' said one of the technicians.

I sang:

A horse is a horse, of course, of course
And no one can talk to a horse, of course
That is of course unless the horse is the famous Mr Ed!

Go right to the source and ask the horse
He'll give you the answer that you'll endorse
He's always on a steady course – talk to Mr Ed!

The technicians joined in:

People yackity-yak a streak and waste your time of day
But Mr Ed will never speak unless he has something to say!

A horse is a horse, of course, of course
And this one will talk till his voice is hoarse
You've never heard of a talking horse? Well listen to this:
'I am Mr Ed.'

Edward clapped us.

'Needless to say Mr Ed isn't you,' I said. 'Mr Ed is that.' I waved my hand at that region of the sky. 'Anyone can talk *to* a horse. The miracle doesn't happen till the horse talks back.'

'Mr Ed,' said Edward, musing, and looked up and out into the dark.

The message protocols

Since we needed to talk to 1956 we chose to talk in Morse. A laser with a shutter could send long and short flashes of light into the wormhole. If the dots and dashs were slowed or accelerated they would still keep their relative lengths, the dots short, and the dashs long. We needed to allow for reasonable time to get the attention of whoever was looking, so needed a nonsense signal, something of no import, just a pattern of flashing that might strike anyone who spotted it as the result of design rather than accident.

And, then, we needed a message.

We couldn't say 'Hello from 2009,' or 'Watch out for Cuba,' or 'Stop carbon emissions now.' The message should be as harmless as possible. Something a person might make note of, but not be able to get anyone to take *terribly seriously*. Something trivial, particular, personal and obscure.

That was the protocol: trivial, particular, personal, obscure.

The arguments we had, the questions that went round and round. How could we know what it was safest to do when we didn't know how time travel might work? We couldn't know whether we were dealing with a self-consistent universe, or multiple parallel universes. We spent hours on the problem of whether the act of selecting one message from among several proposed would adversely affect the outcome of our experiment. Was it more dangerous to make an informed choice? Was it dangerous to follow a hunch, have a preference?

'You think then that we should just pick a message out of a hat?' said someone finally, in exasperation.

'Why didn't I think of that – the hat method,' said someone else, sarcastic.

Edward said, 'We just have to do something. There's no sign yet that we *have* done anything. If we have done something we might hear back from ourselves any day now. But maybe we couldn't make up our minds. Perhaps we were deadlocked. Do you want that to be what happened?'

It was Edward's hat we used, a Red Sox baseball cap.

It was my message that was pulled out of it. The address 'particular' and 'historically insignificant', the message 'personal' and 'obscure'.

In morse:

Andrew McAllister
Kingans Ford
R D 1
Mahoe
New Zealand

Then:

-.. --- -. -   .-.. .. ... - . -.   -.. --- -. -   .-.. --- ...- .

The Andrews McAllister

The business about the tradition of 'Andrews' in my grandfather's family came up only once that I remember. It was when my wife was pregnant with our first child. We were having lunch with Mum and Dad. Mark wasn't there. Mark and conversation were mutually exclusive. Mum and Dad had started inviting their sons separately sometime before my wedding. At first it was a purely practical decision. We had to discuss wedding plans, and not get sidetracked. What we used to do was this: the real talk would take place over the food preparation and dishes between Mum and me, or Dad and me, or Mum and my fiancée. The real talk would escape while whoever was left at the table with Mark acted as a silent decoy, drawing his talk.

So, it was just the four of us at that lunch. We were discussing names for the baby. Dad said, 'I suppose that if it's a boy you might call him Andrew.'

I was quite taken aback. I hadn't expected Dad to suggest anything so quaint. 'Why would I call him after myself?'

'It's actually a family tradition,' Dad said. 'There was your uncle Andrew and Granddad Andrew, and other Andrews further back.' He and Mum exchanged a look. 'But we messed it up.'

'How? I am Andrew,' I said, puzzled.

'Yes. *You* are.'

Then my wife said that she couldn't possibly cope with two Andrews, she'd feel like Dr Seuss's Mrs McCave, 'Who had 23 sons and named them all Dave.'

'We like Hugh and Bruno,' I said. 'And Bridget for a girl.'

It turned out that Mum really disliked the name Bridget and the conversation took a turn away from the messed up tradition and never returned to it.

The boy in the background

I want to say that my message wasn't a result of years of brooding. It did conform to our protocols, and so did the other five I came up with. It was, however, the last one I thought of, and was the result of an impulse.

At the time we were finishing our work on the Photon Telegraph, and thinking up messages, I was alone in New Mexico. My daughter was at university in New York, and my son had gone back to New Zealand with his mother. He wanted to study veterinary science and had decided to do it in Palmerston North. My wife went with him to help look for accommodation and buy him toasters and duvets and whatever else he needed.

I found being alone dreary and depressing. One night I happened to be eating out of tinfoil and grazing channels and I found myself watching a documentary.

It was about autism. I didn't have a revelation about Mark – I'd pretty much worked out for myself what Mark's problem was years before. I'd made my own informal diagnosis. But there was something in the documentary that set me off thinking about myself.

In one scene the farmer with an 11-year-old autistic son was letting the boy drive his tractor. 'It's something he likes to do,' the farmer explained. He was sitting behind his son while the boy held the shuddering steering wheel and laughed. The tractor was jolting over a paddock with cropped grass. The farmer had been feeding out, there were piles of dry hay scattered all over the paddock. The trailer that had held the hay was empty – except for the other boy.

The farmer talked about what he was able to do for his son, how much thought he and his wife put into finding out what the boy enjoyed, and what was good for him. Meanwhile the boy clutched the wheel, his face tilted to the sky, grinning and riding the vibrations.

The other boy, a younger son, the normal one, was sitting in the trailer with his legs folded under him. He was in the background. He wasn't the subject of the shot. He knew he wasn't a subject, and didn't expect to be. The look on his partly averted face, and his pose, both expressed what he was doing, what it was his habit to do. He was being *no trouble*. And he was waiting.

He was waiting, he was waiting, he was waiting.

(Edward's hat, and my message. An address: my name, my uncle's name, my grandfather's name, his father's name, his father's uncle's name. And the message:

    -.. --- -. -   .-.. .. ... - . -.   -.. --- -. -   .-.. --- ...- .)

Internal difference

Last month, my mother finalised the sale of her house. It was sold within the family, but the sale was handled formally by lawyers so that no one would feel precarious or unhappy. I went home to help Mum with the purchase of a spruce one-bedroom villa in a retirement village named after a notable New Zealander who – Mum said – would probably never have dreamed of going near such a place.

My wife was due to join me later and help me sort out the old house, and our family's years of accumulated stuff.

Before I left New Mexico I sent a postcard to Mark. I suggested that he might like to help with the shift.

Mark had been working for the past year as a researcher in a government department. He was earning, and getting out to his office and the supermarket at least. Though his bungalow's exterior paint was flaking and his garden was overgrown, that was nothing new. Houses can look shabby on the outside and attract 'Do you want to sell?' flyers from every real estate agent in a city, but it's what's inside that counts. The insides of houses give away things about their inhabitants' morale and mental health.

It was a still, sunny spring day when I arrived at my brother's house. Mark let me in, and led me to the kitchen. The hallway was a little dank, and had the sour smell I associate with my brother. That too wasn't new or unusual. As we went by the living room Mark pulled its door to. I could hear the gas heater in there, hissing. The sunlight through the kitchen windows showed up a peach fuzz of dust on the front of the cupboards, and the clouds of old spills on the floor, haphazardly wiped. The kitchen was a little cleaner than it had been on my last visit. I looked around me and kind of took the pulse of the house, trying to figure how my brother was, how time was treating him. The kitchen curtains were hanging part way off their drooping curtain rods, and a gap in the window frame had been sealed with masking tape, but the draining board was clean, and there was perhaps only as little as one month's worth of recycling bundled up by the back door.

Mark made me a cup of coffee. My cup had a skin of brown on its inside, but that was only staining, inert now and harmless.

I asked my brother whether he'd got my postcard.

'Yes, I got it.'

'Well, would you like to come down and help?'

'No thank you.'

Right away I realised my mistake. I had asked him whether he'd *like* to help. He used to behave this way about the dishes. At those family lunches, after lunch I'd have my hands in the suds, and I'd say to him, 'Mark, would you like to come and help?' And he'd get a smug little smile on his face, and would reply, 'No thank you.' You see, the joke was on me for being so conventional, so tied up by false politeness that I couldn't just simply say what I wanted. Mark would feel that he was giving me a little rebuke for my little dishonesty, and he wouldn't have to get up from the table.

'We really could do with an extra pair of hands,' I said.

'Andrew, it was your choice to come over and get involved. It was your idea. I don't see that I should let you organise me according to your ideas.'

In my exasperation, I made another mistake. I said, 'Any normal son would help.'

Again I saw the small gleeful light come into his eyes. He said he'd been doing some reading. He pointed to his kitchen table and a pile of books. I could see that they were about what Mark was now calling 'My Syndrome'. He said he'd learned so much about himself. He'd made sense of all these things that used to *happen* to him. Like, when he was young, and he used to go to parties and everybody would be talking to one another, and the room would be filled with noise. 'It was so confusing, and I'd just sit in a corner. I didn't know what to say to anyone, whereas everyone else seemed to know what to say to one another.'

I listened to my brother describe his old social crucifixions and thought – yes – Mark would hear a room full of people talking as noise, because he was unable to choose which person to listen to, which conversation to board and let carry him away. This was his tragedy. But then I thought, 'Mark hears a room full of people talking as noise, and I understand that this is because of the way his brain works, *but . . .*'

But Mark was going on about the 'neurologically normal'. 'The neurologically normal are multi-taskers,' he said. 'Their thinking is fast and flexible. But people like me think *deeply* about only one thing at a time. We're more contemplative.'

Mark's latest thing was himself. He had renewed his narrative. He'd now admit that he was out of step, and easily troubled, but the consolation was that he was contemplative, as well as an honest soul. Apparently I was one of the 'neurologically normal', part of a bunch of yammering, glib multi-taskers: the tribe of apes, social, busy, thick-skinned – according to Mark's relentless solipsism. Now that he understood how he came to be so at odds with the world he was congratulating himself. I thought of the hundreds of occasions on which I had gone out with my brother when we were growing up and spent the better part of my time ministering to his shyness, reassuring him, drawing him out for others to hear.

What a *kind fool* I'd been. Now here was my brother busy with his new hobby. Too busy to take down our mother's pictures, to box her books, to wipe her skirting boards.

I sipped my coffee and let Mark talk. After a time he got up to fetch something from the living room. When he opened the door I saw that there were two clothes horses full of washing drying by the gas fire. I said, 'What is all that doing in there?'

'I don't really like the backyard,' Mark said.

'What's not to like?'

He squinted. 'It's the neighbours.'

'Are they nosy?'

Mark looked shifty. 'It's not exactly that.' Then in a burst, 'You see, there's this developer. He lives four doors along, in the grey house, eh. He's bought up a lot of the properties around here, hasn't he. I think he'd like to have the whole block, wouldn't he. He's some kind of Christian. Maybe he wants to make the neighbourhood into a church compound, eh? Anyway, I've been noticing that, over the past few years, the trees around my boundary are dying slowly. Every year their leaves are smaller and sparser, aren't they. It's suspicious, eh. This guy is poisoning my trees, isn't he. He has tenants in all the properties on my boundary. I just bet they all go to the same church, eh. Anyway, I think he has them doing something to my trees. Sometimes I hear voices in the backyard at night. People are creeping around, eh. I think that developer understands how much I value my privacy, and how easily disturbed I am.'

I felt that I was slipping, falling into the dark. I felt that I was being swallowed and crushed. I rallied myself. 'How long do you think this has been going on?'

'How long has he been poisoning the trees? He's doing it so that it looks as if they are dying back naturally. So he has to make sure he takes a long time about it, doesn't he. I think he hopes I'll just decide I don't like living here any more, and will put the place on the market.'

'But Mark, how does someone poison trees gradually?'

'I don't know, but there must be a way because that's what's happening. And you remember how our walnut tree died from bacteria in the soil? Do you remember how its leaves got wrinkly, and smaller?'

'Mark, I have to ask you, how likely is it that someone is poisoning your trees?'

It seemed that Mark couldn't understand my question. His mind didn't work well with likelihood, with probability. Anything bad that could possibly happen would probably happen to him. Besides, what people did and were able to do – those were imponderables.

My mind sidestepped; it didn't want to deal with any of this. 'All right,' I said. 'If you don't like going into your backyard to hang out the washing because of the neighbours, why don't you get yourself a dryer?'

Mark looked incredulous. 'Don't you know that second only to smoking the biggest cause of house fires is tumble dryers? It's the combination of lint and static electricity,' he explained, then went on for some time about static charges. I listened, and now and then tried to reason with him. But reasoning with Mark always caused him distress and I eventually backed down. I finished my coffee and went away. I just didn't know what to do with him – with this combination of crippling fear and crippling arrogance.

A curse

The dust we raised aggravated my mother's eczema, so we convinced her to keep to the kitchen, or potting shed, where she was striking cuttings from her favourite shrubs, far too many for her small garden at the retirement village.

There were a lot of papers. It seemed that my parents had kept all their correspondence, even 25-year-old electricity bills. (That would be my father, always ready to be called to account for all the little things in life, his fear of disorder so acute perhaps because he felt he'd failed with Mark – as the father of a child beyond the help of mere good parenting, close attention, love even.)

My wife and I sorted, we rolled up our sleeves and got stuck in.

A day or two into the task my wife was sitting at the dining-room table with a suitcase full of old correspondence in front of her. We had decided that it was best to go through discarding all the old envelopes first, before reading the letters. For some time my wife had been doing this, making a soothing series of crackles and rustles, opening envelopes, freeing sheets of paper, smoothing them flat.

I was sorting Dad's tools into those for Cash Converters, those for the Salvation Army, and those fit only for a trip to the tip. After a time I became aware that the rustling at the table had stopped. There was a silence in the room so intense that it seemed to vibrate against my lowered eyelids. I looked up at my wife. She had a blue aerogramme in her hand. She was gazing at it, but a second after I looked at her she raised her head and returned my look. Her face was pale.

She got up, came over, and handed me the aerogramme.

It was addressed to Andrew McAllister, Kingans Ford, RD 1, Mahoe, New Zealand. It was postmarked San Diego, 10 June 1956.

I opened it. The blue rice paper was fragile, furry, and split along its folds. Inside were only four words, scrawled, either hurried or impulsive.

'Don't listen. Don't love.'

The letter had a return address, and the sender's name was that of the astronomer at the Palomar Observatory, the one man who had stubbornly refused to relent on the matter of the 'comet from nowhere'.

A green light in the sky

Of course we knew where he was to be found. Almost since Eric Hall's visit to Edward at Amherst we had known where he was.

Because it had been decided to keep the salient fact about The Deity secret – the location in time of the downstream end of the wormhole – we had been vigilant in keeping a watch on anyone who might work it out for themselves. But it seemed that 'the comet from nowhere' was a matter best forgotten. Almost everyone involved had gone on assiduously to live it down. We did keep an eye on the one man who'd persisted in talking about it. Would he figure it out? Would he publish something? Would he talk to the press?

He was an old man already six years back. He and his wife had retired to one of those small towns in Mexico that frugal Californian retirees began to discover some time in the 1990s. We found him there and watched him. But he never seemed to take any interest in the possibility of a connection between his old *bête noire* and the disappearance of Ranchod's Comet. He played golf, and went fishing, and was silent to the best of our knowledge.

How *naïve* we were, keeping an eye on the old man, with no thought of what action we'd take if he did submit an article or talk to a reporter. All we wanted was some kind of forewarning. We thought about what kind of spin we could put on our secrecy.

How terribly naïve we were. Other interests were involved. There were the people who funded our experiments. We had been sworn to secrecy. Why couldn't we imagine that those people would make sure that secrecy was preserved? For they too had been watching that old man. They kept an ear in his house in Mexico, they observed what he did and who he met, and they too were reassured that he'd put the past behind him, and his former life as a truth-telling scientist beset by doubters. They watched him enjoy his old age, his second wife, his grandchildren, his motorboat, his barbecue, his golf.

But, of course, they were *still* watching when, a week ago, Edward and I visited him at his last address, a nursing home in San Diego.

He was in a wheelchair in the quiet day room, the one without a television. A nurse conducted us to him, and found us some chairs. He looked pleased to have visitors, and yanked the earphones of his iPod out of his furry ear holes. He forgot to turn it off though, and the whole time we talked I could hear its faint insectile singing.

Edward and I introduced ourselves, and shook the cold bundle of bones that was his hand.

Ed asked him how old he was.

Ninety-two next birthday.

'So, you were in your 30s in the 50s?' Edward said. We already knew the answer, and it occurred to me that Ed was testing the old man's mental sharpness.

'Yes. My children were all born in the 50s,' said the old man.

Ed settled back in his chair, smiled in satisfaction, and signalled to me by lifting his eyebrows.

I gave the old man the aerogramme.

He gazed at it for a long time, then said, 'Oh.' He peered hard at me. 'Did I hear you say that your name was Andrew McAllister?'

'Yes.'

'Oh,' he said again.

'It was my uncle's name too, and my grandfather's,' I said.

'Ah.'

Ed asked, 'Can you tell us about the signal?'

The old man carefully folded the letter and gave it back to me. He began, 'It was late April that year – 1956. I could have given you the date once. I had the pages from the Hale logbook up until – well – whenever it was that my daughter and her husband packed up my house in Mexico. I'm afraid I couldn't say now where those pages have got to.'

'But there were no pages missing from the logbook,' I said.

'We tore out two leaves and just rewrote it, rewrote everything that happened over those nights, but without the green light. The book wasn't printed up with dates. You see, back then we used to do everything by hand, there was none of this bloody *template* stuff. I hate the way my computer is always telling me how to do things.'

'Tell us about the green light.'

He drew himself up, straightened his spine as much as he was able. He gave us a position, a region of the sky. 'It started up around 3 a.m. A flashing green light. We thought it was an aircraft at first. But it wasn't. It was coming from out there, from the edge of our solar system.' He paused and narrowed his eyes. 'I suppose you two gentlemen must have something to do with that wormhole, The Deity. I suppose you're part of that outfit in New Mexico.'

'Yes,' Edward said.

'Then you must know the flashing was Morse. Two dots, a long space, a dot and a dash, a shorter space, then two dashes. It went on like that for close to an hour.'

Ed and I exchanged a look.

'It was saying, "I am, I am, I am", over and over again. You can imagine what we felt.'

We nodded.

'Though we couldn't imagine why God would choose to talk to us in Morse. But then it suddenly changed. It said something else, and only once. I wonder if you can guess what it said?'

'The name and address. The contents of the letter.'

'No. That came later. The following evening at nine or so that started up and went on for several hours.'

'Okay, we give up, what did it say on the first night?' I asked.

The old man leaned forward. He whispered, 'I am Mr Ed.'

Those bloody technicians.

'That television programme, *Mr Ed*, didn't start till the 60s, thank goodness. We weren't forced to try to make sense of any silly talking horse. Later, when it did come on, it brought it all up again for me.' He shook his head. 'It was a terrible time,' he said. 'A terrible time.'

'But that night,' Edward said, prompting. 'Summer, 1956 . . . ?'

'We thought it had to be a hoax, but couldn't work out how we were being hoaxed. Or why. And this was after the comet. The address and message came in. Flash, flash, flash, for hours. It was like something from the end of the world. It was incoherent. We could read it, but we couldn't make anything of it. It drove us crazy. We were at one another's throats. We never did decide what to do, we just started doing things. Someone tore the pages out of the logbook. I rescued them from the trash later. Someone else sat down and forged two nights with no green light. After that I just went home and stayed there. I wasn't well. I couldn't get warm. My wife was angry that I wouldn't talk to her. She went off to her mother's and took the kids. I suppose I had a kind of breakdown.'

'But you still sent the letter.'

'I had to. I felt it was a test. I wrote out what the light had said and I put it in the mailbox.'

'But you did keep talking about the comet.'

'Later, years later. I suppose it looks to you, all these years on, as though everything happened within a short space of time. It didn't. Eventually I just couldn't stand the other secret – the green light – so I started to make noises about the comet.'

'The green light was a laser, and when it went into the wormhole it was red,' Edward said.

'You don't say.' The old man mused. 'I wonder how that works?'

Edward said, 'It'll be something to do with gravity.'

For a long moment we all pondered the wormhole's effects on light. Then, between us, Edward and I told him everything else. We told him about Ranchod's Comet, our Photon Telegraph, our message protocols, our six years of dissembling to the world. And then I told him about my grandmother, her husband's family tradition, and about my deafening and insensible brother.

'My parents got your letter when they were expecting my brother. They had planned to call him Andrew. The first-born son of each generation of my father's family had been an Andrew. When I found the letter among my mother's papers I asked her about it. She said that the message, "Don't listen. Don't love", had seemed like some kind of curse. It was like the presence of the uninvited wicked fairy at Sleeping Beauty's christening, she said. She said that my father wrote back to the return address. He sent a letter that more or less said just, "What the hell?" But they never got a reply. My uncle Andrew wrote too. He wasn't living anywhere near Mahoe, but it was supposed the letter might be for him.'

'I couldn't explain it,' the old man said. 'I'm sorry. I did get those letters. But what could I say in my own defence?'

'Why put a return address?'

'I wanted an explanation! But all I got was puzzled indignation.'

'My parents didn't name their first son Andrew. Grandma was very upset. Mum and Dad couldn't explain to her why they'd changed their minds. They didn't want to tell her about the letter, they wanted to spare her that. And they wanted to lift the curse. So they called their first-born Mark. But then he turned out unloving, and unable to listen.'

The old man said, 'It wasn't a curse.'

'I know. It was me trying to warn myself about my brother.'

The old man nodded. 'Thank you. Thank you for telling me all this,' he said. 'I'm glad to know finally, even if it is a sad story. It's like those stories about some king sending to the oracle of Delphi to find out what will happen when his son is born.'

'Oedipus,' Edward said. 'The oracle says that the baby will grow up to kill his father and marry his mother. So the king has his son left out on a mountainside to die. But a shepherd finds the baby and raises him as his own. And so on and so forth. What happens in the end is what the gods always said would happen.'

'Man proposes; God disposes,' said the old man.

'Exactly,' said Edward. And they sighed together like two old men.

But I was thinking that my grandmother on her deathbed *had* blessed me. 'Andrew, I'm glad you're Andrew,' she said. She said, in effect, 'I'm glad it was *you* who got the family name, my beloved husband's name. For perhaps you can be trusted to be a steward of all I care for.'

You've never heard of a talking horse?

Today I'm packing up my office, my personal effects, under the eye of a man in uniform.

Yesterday they came and closed the lab – to us at least. They sealed and removed our filing cabinets and our computers.

Tamara was the only one of us who would always ask why we'd found no sign of ourselves in the past, in the historical record. No sign of our own characteristic actions, the sorts of things *we* might be likely to decide to do. She used to say that if we hadn't heard from ourselves, hadn't recognised our own doings, it would mean that we were living in another world than the one altered by our interference.

But now we know that we haven't heard from ourselves because it was taken out of our hands – the project, the Photon Telegraph, the open ear of The Deity itself.

And that is the story so far, a history of time travel's first ill-fated petition. It is out of our hands. And that is why you've never heard of our talking horse.

=

GLENN COLQUHOUN

WITH TONY SIGNAL

The theorem of Pythagoras

$$a^2 + b^2 = c^2$$

In a right-angled triangle the square of the hypotenuse is equal to the sum of the squares of the other two sides.

a = the altitude of a triangle
b = the base of a triangle
c = the hypotenuse of a triangle

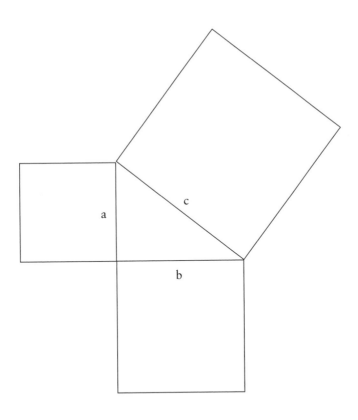

$$a^2 + b^2 = c^2$$

In the local supermarket
a man travels north

between the canned
goods and the sugar.

In his basket:

skin,
bone,
doubt,

mayonnaise
and
candles.

His heart is a small red tin.

A woman passes west
to test the fruit.

In her trolley:

potatoes,
matches,
lipstick,

loneliness
and
danger.

Between these two
a shared hypotenuse:

flesh and blood,

hope,
potato salad
and candlelight.

The law of reflection

$$i = r$$

The angle of reflection is equal to the angle of incidence.

i = the angle of incidence
r = the angle of reflection

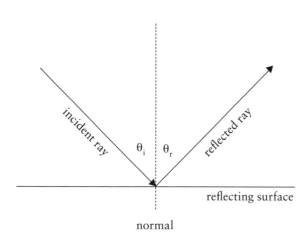

i = r

The angels call   Knock-knock-knock-knock-knock.
at half past four.   Knock-knock-knock.
                         Knock-knock
They bring,   and knock.
they bring
the light.   At the door
                         they stand.
They leave   Behold!
their bikes
at the gate.   The light
                         they bring,
They pass   they bring.
the letterbox,
the fig,   They pass
the lemon tree,   the pot plants,

the cat   the dog,
on the grass,
                         the lawnmower
the car   on the lawn,
in the driveway,
                         the hose
the hose   in the driveway,
on the lawn.
                         the car
The lawnmower,   on the grass.
the dog,
the pot plants –   The cat,
they pass.   the lemon tree,
                         the fig,
They bring,   the letterbox,
                         they pass.
they bring
the light.   At the gate –
                         their bikes.
Behold!
                         They leave.
They stand
at the door   The light
and knock.   they bring,
                         they bring
Knock-knock.   at half past four.
Knock-knock-knock.
Knock-knock-knock-knock-knock.   The angels call.

Newton's three laws of motion

The first law:

$$\mathbf{F} = 0 \Leftrightarrow \mathbf{v} = \text{constant}$$

Every object continues in its state of rest, or of uniform motion in a straight line, unless it is compelled to change that state by forces impressed upon it.

$$\mathbf{F} = \text{force}$$
$$\mathbf{v} = \text{velocity}$$

The second law:

$$\sum \mathbf{F} = m\mathbf{a}$$

The acceleration of an object is directly proportional to the net force acting on the object, is in the direction of the net force, and is inversely proportional to the mass of the object.

$$\mathbf{a} = \text{acceleration}$$
$$m = \text{mass}$$

The third law:

$$\mathbf{F}_{12} = -\mathbf{F}_{21}$$

To every action there is an equal and opposite reaction.

$$\mathbf{F}_{12} = \text{force of object 1 on object 2}$$
$$\mathbf{F}_{21} = \text{force of object 2 on object 1}$$

$$F = 0 \Leftrightarrow v = \text{constant}$$

There was an old woman from Mercer
whose husband was sick of inertia.
He cried with a thump,
'I'll give you big lumps,
I married one, now I could curse her.'

$$\sum F = m\mathbf{a}$$

The nun to the priest in the sack
said, 'My dear it's Newtonian maths.
Although I don't rate
how you accelerate,
I'm greatly impressed by your mass.'

$$\mathbf{F}_{12} = -\mathbf{F}_{21}$$

A large man from Haast was aghast
when his wife kicked him hard in the arse.
She fell with a whack.
He said, 'Dear, I kick back,
but I sometimes go forward when I fart.'

Newton's law of universal gravitation

$$F = GMm/d^2$$

Every particle in the universe attracts every other particle in the universe by a force directed along the line connecting the two. This force is proportional to the product of the masses and inversely proportional to the square of the distance between them.

F = gravitational force
G = Newton's gravitational constant
M = the mass of particle 1
m = the mass of particle 2
d = the distance between particles 1 & 2

$F = GMm/d^2$

Falling is a space, he said;
The apple to the earth inclined.
I feel the embrace, she said.

All the stars above our heads
Fill the sky with purple lines.
Falling is a space, he said.

You and I are so ensnared;
One to fall and one to climb.
I feel the embrace, she said.

Draw me down into your bed,
Hold me to your lips like wine.
Falling is a space, he said.

All the planets overhead
God within his arms aligns.
I feel the embrace, she said.

But I will orbit you instead
And you will be my only time.
Falling is a space, he said.
I feel the embrace, she said.

The laws of thermodynamics

The first law:

$$\Delta U = Q - W$$

When a system undergoes change, the change in energy equals the heat transferred to the system minus the work done by the system, i.e. heat is another form of energy.

U = the internal energy of a system
Q = heat transfer
W = work done by the system

The second law:

$$\Delta S \geq 0$$

The total entropy of an isolated system that undergoes a change cannot decrease, i.e. order tends towards disorder.

S = entropy

The third law:

$$T > 0$$

All processes cease as temperature approaches absolute zero, i.e. absolute zero can never be reached.

T = absolute zero

THE LAWS OF THERMODYNAMICS

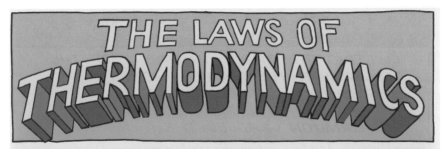

$$\triangle U = Q - W$$

KNOWING IN A MATTER OF SECONDS THE PLASMATRON EXCHANGER WOULD REDUCE METROPOLIS TO RUBBLE, **QUARK KENT** STEPPED INTO THE RED TELEPHONE BOX— GLAZED WITH RAINWATER.

MOMENTS LATER...

A BLINDING *FLASH* THE ROLL OF DISTANT THUNDER...

A PAIR OF FAMILIAR BLACK GLASSES STARING UPWARDS FROM THE ASTONISHED PAVEMENT.

△ S ⩾ ○

CUTTING THE TRACK IN FRONT OF THE TRAIN WITH HIS X-RAY VISION THEN CAREFULLY CALCULATING THE PARABOLA OF THE FALLING CARRIAGES, **SUPERMUON** CRADLED THE RUNAWAY LOCOMOTIVE TO THE GROUND.

MEANWHILE...
        ACROSS TOWN...
CHAOS SPREAD QUICKLY THROUGH THE CITY. OVERWHELMING THE GUARDS WITH THEIR GUARD-OVERWHELMING MACHINES A CLUTCH OF CRIMINALS ESCAPED FROM THE STATE PENITENTIARY.

SUPERMUON MUST DIE. SUPERMUON MUST DIE.

HA-HA-HA-HA-HA!
HA-HA-HA-HA-HA!
HA-HA-HA-HA-HA!

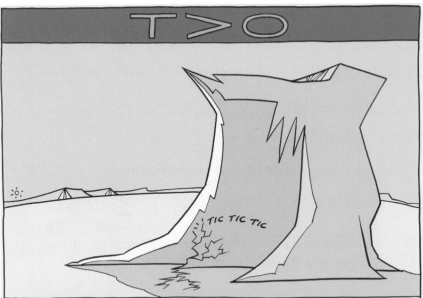

TRAPPED IN THE FRIGID EMBRACE OF A
DISTANT ANTARCTIC GLACIER DEEP BELOW TWO
HUNDRED TONS OF LEPTONITE. SUPERMUON'S
BODY LAY HELPLESS. THE ONLY SOUND A
STRANGE CLICKING DEEP BENEATH THE TUNDRA.

TICK... TICK... TICK... TICK

A CHILLING REQUIEM?
A SUPERHEART STILL BEATING?
A SMALL CRACK OPENING
IN THE ICE?

Maxwell's equations

1. Gauss's law for electricity:

$$\nabla \cdot \mathbf{E} = \rho / \varepsilon$$

The electric flux of a closed surface is proportional to its charge.

2. Gauss's law for magnetism:

$$\nabla \cdot \mathbf{H} = 0$$

The net magnetic flux of a closed surface is zero, i.e. all magnets must have both a north pole and a south pole.

3. Faraday's law of induction:

$$\nabla \times \mathbf{E} = -\mu \partial \mathbf{H} / \partial t$$

A changing magnetic field produces an electric current.

4. Ampere's law:

$$\nabla \times \mathbf{H} = \mathbf{J} + \varepsilon \, \partial \mathbf{E} / \partial t$$

A changing electric field produces a magnetic field.

$\mathbf{E}$ = electric field strength
ρ = charge density
ε = permittivity
$\mathbf{H}$ = magnetic field strength
μ = permeability
$\mathbf{J}$ = current density
t = time

$$\nabla \cdot \mathbf{E} = \rho / \varepsilon$$

Room full of men,
a woman's voice –
all heads turn.

$$\nabla \cdot \mathbf{H} = 0$$

Heron to the north,
godwit to the south –
nothing changes.

$$\nabla \times \mathbf{E} = -\mu \partial \mathbf{H} / \partial t$$

Flight of birds,
all heads turn –
a woman's voice.

$$\nabla \times \mathbf{H} = \mathbf{J} + \varepsilon \, \partial \mathbf{E} / \partial t$$

A woman's voice,
all heads turn –
heron, godwit.

Einstein's mass-energy equation

$$E = mc^2$$

The rest energy of an object is its mass times the speed of light squared. Mass is a form of energy.

E = energy
m = mass
c = the speed of light

$$E = mc^2$$

In No. 3 an old
woman knits –

at her feet

twelve pieces of string,
a fall of rain,
the cat's eyelash.

She makes socks,
a hat, her
neighbour's coat.

Next door,
at No. 4,
her sister unpicks –

socks, a hat, her
neighbour's coat.

At her feet

twelve pieces of string,
a fall of rain,
the cat's eyelash.

Outside,

tall clouds billow –

an old, grey man

staring down on his
argumentative
children.

General relativity

$$G_{\mu\nu} = 8\pi G T_{\mu\nu} + \Lambda g_{\mu\nu}$$

The curvature of space-time is determined by the distribution of mass–energy.

$G_{\mu\nu}$ = the Einstein tensor
G = Newton's gravitation constant
$T_{\mu\nu}$ = the energy–momentum tensor
Λ = the cosmological constant
$g_{\mu\nu}$ = the metric tensor

$$G_{\mu\nu} = 8\pi G T_{\mu\nu} + \Lambda g_{\mu\nu}$$

If we pick up a stone and then let it go, why does it fall to the ground? The usual answer to this question is because it is attracted by the earth. Modern physics formulates the answer differently. More careful study of electromagnetic phenomena, has led us to regard action at a distance as a process impossible without the intervention of some intermediary medium. The action of the earth on the stone takes place indirectly. The earth produces in its surrounding a gravitational field, which acts on the stone and produces its motion of fall. As we know from experience, the intensity of this action on a body diminishes as we proceed farther and farther away from the earth. This means the law governing the properties of the gravitational field in space must be seen as definite; **falling is a** response to **place**. Due to its location alone, one body is able to influence ano**ther**. It is usually **said** that this attraction between the planets is a distinct force which exists independently of the bodies it affects, however the universe is not **hard-edged**. People, **streets**, priests **and church**es, men and women create similar fields, even **a man walking a dog**, according to the principle of relativity, can be viewed alternatively as a dog walking a man. I am sure some people will express surprise. **What is the** effect of this? To **address** accurately the technical applications of these idea**s here** it must be **said** that other forces are known to exist which also distribute their influences in a similar manner. In an electromagnetic field an electric force marches **first right then left** in tandem with a magnetic force. Magnetic forces likewise step left **then right then left** in tandem with electric forces. Gravitational fields are similar, however they exhibit another important property. It i s thought that they exist **somewhere on the skin** of t**he said** universe. The effects of this are most apparent **where the surface** of space-time **bends** due to the presence of matter; a solid universe accelerates bodies along a geometry created inside it by mass. To reiterate clearly, **falling is** the response of **a** body **place**d in relation to another. Stated in another way, **falling is a** dis**place**ment of space-time by two or more objects within it. How is this accompli**she**d? Simply **said**, we are led to the conviction that, according to the general principle of relativity, the space-time continuum cannot be regarded as a Euclidean one. In gravitational fields there are no such things as rigid bodies with Euclidean properties; thus the fictitious rigid body of reference is of no avail in the general theory of relativity. According to Newton's law of motion *force = inertial mass x acceleration* where the 'inertial mass' is a constant of the accelerated body. If gravitation is the cause of the acceleration, we have *force = gravitational mass x intensity of the gravitational field* where the 'gravitational mass' is likewise a constant for the body. If the acceleration is independent of the body and always the same for a given gravitational field, then the ratio of the gravitational to the inertial mass must be the same. The gravitational mass of a body is equal to its inertial law. If the reader has followed our previous considerations, he will have no further difficulty in understanding all the methods leading to the solution of the problem of gravitation.*

* Adapted from *Relativity: the Special and General theory* by Albert Einstein.

Schrödinger's equation

$$i\hbar \, \partial\Psi/\partial t = \hat{H}\Psi$$

The Hamiltonian operator determines the time evolution of a system. This equation represents matter as a set of probability waves. It shows how these waves alter under external forces. Its solution gives the probability of finding any particular particle at any particular time in any particular location.

$\hbar$ = Planck's constant
Ψ = the wave function of a system
$\hat{H}$ = the Hamiltonian operator
t = time

$i\hbar \, \partial\Psi/\partial t = \hat{H}\Psi$

At Dorothy's but not at Dave's
Everything was made of waves.

The chimney curled.
The kitchen leant.
The curtains bulged.
The windows bent.

At dear old Dave's but not at Dot's
Everything was made of spots.

Dots. Spots.
Lots. Blocks.
Lots of spots in
dots in blocks.

Once Dave tried to live with Dot
But he found that he could not.
When he sat down in her chair
He found it wasn't always there.

When Dot tried to live with Dave
Very soon she grew dismayed.
Deep inside the bath she hopped
To find the water made of dots.

'I do not like your dotty ways,'
said Dorothy to dear old Dave.

'I do not like your lumpy waves,
I feel sick,' said dear old Dave.

Dorothy, to keep the peace,
Cut her waves up piece by piece.
Unhappy with the status quo,
Dave tied his dots into a bow.

'It's true. It's true,'
said Dorothy,
'Every one is really two.'

'Ho hum. Ho hum,'
said dear old Dave,
'All my dots are really one.'

They may not be the perfect match
But they will do for now perhaps.
They see each other for all that
When they both put out the cat.

At Dorothy's but not at Dave's
Everything was made of waves.

At dear old Dave's but not at Dot's
Everything was made of spots.

The Yang-Mills Lagrangian
for quantum chromodynamics

$$L = \bar{\psi}(i\gamma^{\mu}D_{\mu} - m)\psi - G_{\mu\nu}G^{\mu\nu}/4$$

The nuclei of all atoms are formed from the interactions of quarks and gluons. These interactions are mediated by the strong nuclear force. This equation describes that relationship.

L = Lagrangian
$\bar{\psi}$ = the quark field
i = the square root of −1
γ^{μ} = the Dirac matrices
D_{μ} = the covariant derivative
$= \partial_{\mu} + igA_{\mu}$
g = the coupling constant
A_{μ} = the gluon potential
m = the quark mass
$G_{\mu\nu}$ = the gluon field tensor
$= D_{\mu}A_{\nu} - D_{\nu}A_{\mu}$

$\bar{\psi}$

What if I thought of you as bees?

Those small black dots.

Perhaps there are things finer
but these will do for now.

They meet the specific criteria.

They are born. They die.
They live in fields.

When the sun assumes
the perfect bloom
the long grass throbs:

'To bee or not to bee.'
'To bee or not to bee.'

This seems reasonable somehow.

i

But who on Earth are you?
I expected a drone or two,
three or four workers at most.
Not this strange, black swarm drifting
through flowerbeds like a drunk
man heaving into rubbish bins.

It's not enough to wave the corn
and bend the trees, you have
to upset the bees as well.

Now they are reliable as the sea.

γ^μ

Damn you light!
Get out of my eyes.
Can't you see I'm busy
taking down particulars?

The length of wings.
The angles they make to the body.
The accumulation of legs
and eyes and stings.

You remind me of a tax collector
constantly making adjustments.

Right now, staggering under
the trees like a new-born foal,
you do not seem so tough.

Who would have thought
all height and depth
and width and time would
one day bend to you?

D_μ

Some bees are fast.
Some bees are slow.
Some bees are in between I know.

g

Some bees kiss.
Some bees don't.
Some bees will and
some bees won't.

A_μ

Now we arrive at the heart,
this, at the centre of the swarm.

The long-stemmed,
colour-drizzled,
mirror-petalled flower.

That thing the bee loves most.

That thing approached
from north and south
and east and west which always
remains exactly the same.

Our firm foundation.
The sure footing.

That thing inside the thing inside
the thing inside which
there is nothing else to find,

except of course for nectar.

How often beauty is the last one
to see the secret cause alive
as though symmetry itself
might somehow be found
with the blood-soaked
wrench lying under its bed.

m

I won't forget:

some bees are big,
some bees are small,
some bees are short,
some bees are tall.

$G_{\mu\nu}$

None of this would work,
of course, without honey.

To begin with, it is sticky.

This is a fundamental requirement.

Further to that, it holds everyone together.
What else would bees do all day?
Lie around picking fights?

Thirdly, it begins inside the flower,

dangling precariously
at one end of that
narrow tunnel between
men and women.

Damp and sweet.

This is a favourite haunt of bees.

Last, but not least,
it is a sensible roof over
every head at night –

a hub, a pub, the footy club.

I would have thought it was obvious.
If bees have a god,
he would be made of honey.

Come to think of it he'd be
useless without them.

$G^{\mu\nu}$

Six-sided wonder.
Small house of wax.
Is it true?
Did honey and the bee make you?

Empty roses,
hunger and desire,
how ridiculous
that we are made of this!

¼

And just when I'd begun to swallow
my own bullshit you show up,
telling everyone to believe a fraction at most.

I'd have thought that was generous.

You give me no quarter.

I am divided in half and
then in half again.

$$L = \bar{\psi}(i\gamma^{\mu} D_{\mu} - m)\psi - G_{\mu\nu} G^{\mu\nu} /4$$

And so in the end
all I can say for sure
is that solid ground
is an empty field.

Inside the field
is a swarm of bees.

Inside the swarm
is a perfect rose.

Inside the rose
is a simple kiss.

And everything I feel
And everything I love
And everything I am

is made from the
irresistible attraction
of the bee to the flower.

THE DEADLIFT

CATHERINE CHIDGEY

There was a time when I was the strongest boy in the world. I have sashes emblazoned with that title; I have trophies and certificates and shields. I remember the way my body felt: hard, defined, a carapace of muscle. At the lab, the calipers scrabbled at my skin like chopsticks and found not a morsel too much, but even then, I saw room for improvement. I knew all the places a body could hide its flaws. My mother told me that before I was born I punched and stretched and kicked until she was bruised inside, until her heart was black and blue. I crawled at six months and walked at nine. And at this time, when I was a creature without words, when the top of my skull was still soft and open to suggestion, she began to exercise me. She lay me on the bunny blanket and moved my little limbs up and down, up and down, pumping my arms as if saving my life. As soon as I could crawl, she tied weighted bags to my wrists and ankles so that every movement was useful, every random action part of my instruction. She calculated how much I could bear, and then encouraged me to bear more. Dr Sime is quite correct when she notes that I did not initiate this regimen, but she overlooks the fact that by the time I could talk, I never once refused a training session. On the contrary, I looked forward to those hours, when my mother and I worked as one body again, as if we had never been parted. She wanted me to be happy, and she was always checking my mood, as if the slightest change might signal cause for concern. She always said I must tell her if I was no longer enjoying myself.

After two years of entering the smaller, local competitions, I won my first title at the age of four, and from then on we took things seriously. Baby Atlas, Junior Muscles, Little Mr Hercules: I won them all. My mother quit her job at the institute and became my manager and agent, and we were often away from home for weeks at a time while I competed all over the country. I loved travelling, watching my home town drop away as the plane took off and rose above the clouds, where the sky was always perfect. I would settle into my seat and listen to music on my Walkman or play Pacman on my mother's calculator. I was good at that; I could gobble up legions of aliens as quickly as vitamin pills.

I don't tan easily – I imagine this comes from my father – but for competitions the skin must be as dark as teak, so that the muscles reveal themselves, so that they cast a shadow. My mother knew how to colour my skin so they stood out. Before each meet she applied the fake tan to my body, taking care to massage it into my hairline and to cover my palms and the soles of my feet. I became a different person, my pale eyes stark against my face. I learned not to flinch when she stroked the backs of my knees and under my arms; I was not ticklish, not like other children who shrieked and screamed and ran from their parents. When she rubbed the stain into my chest, I felt calmed, and when she did the whorls of my ears, I heard the ocean. In the shower the fake tan turned the water brackish. It pooled at my feet, spattered the white walls, as if there had been a storm inside. I scrubbed at my skin until the water ran clear, but even then a layer of colour lingered for a time, so I was never sure what my real complexion was. Every day, with my breakfast, my mother gave me vitamins from a bottle with no label.

I had my rivals, of course; the prizes were such that there were always new contenders. My mother taught me not to talk to them, but to concentrate instead on my stance and my lifts, running through my routine in my head so that when my name was called I could walk onto the stage and perform without thinking. *Smile*, she willed from the audience, stretching the sides of her mouth apart as wide as they would go, showing all her white teeth. The other mothers always made a space for her.

I was not like the other boys my age, and I was not like their elder brothers either. At breakfast I did not sit down to a plate of cereal or toast; I did not drink glasses of orange cordial or mugs of hot chocolate. My mother made me liquid meals in the blender instead: bananas and berries and milk, and of course the powdered protein supplements that always left a sandy residue in the bottom of the cup, as if someone had used it for sandcastles. In the school playground I danced across the monkey bars; I slipped like water through the cool concrete pipes. I twisted the chain handles of the swing, winding myself up then letting myself go so that I spun and spun, and the playground disappeared, and all the other, weaker children disappeared. When we were learning about the solar system Mr Mayhew took us outside late in the afternoon, just before the bell rang. He made us stand very still while he traced our shadows with chalk. I could smell the stones and the dirt and the stale sandwiches as I stood waiting for my turn. When he had finished there was a forest of bodies on the warm concrete, as if there had been a slaying. They were gone the next day, these chalk children, washed away by rain in the night, but I could not forget my stocky shadow, so different from the others.

My mother and I did well. I was invited to open gyms and health-food shops; my image sold breakfast cereals and long-life batteries and super-strength glue; a line of sportswear for children bore my name. We bought a much bigger house which had its own gym and swimming pool just for me. My mother let me unlock the front door the day we moved in, as if it were another official opening, but I did not rush ahead of her. I turned and scooped her up in my arms and we entered together. She felt as light as a shadow to me.

I was happy in the new house. It had everything I wanted, and I imagined myself staying there forever, like the people who shut themselves in giant glass domes as experiments. There was a wall-sized TV in front of the treadmill and the rowing machine and the stationary bicycle, so that I could pretend that it was a picture window, and that I was outside as I trained, but of course, no matter what distance I covered on the machines, I never ended up anywhere different.

We met The Cougar soon after we moved there. I had seen him at the competitions, prowling about in the halls and auditoriums, a sleek figure with a powerful torso and the bowed, wishbone legs of a serious lifter. He watched the junior competitors as if he would like to devour them, now and then approaching close enough to hand over a business card. These days a man like that would be quietly taken aside and questioned, but back then the mothers were delighted if someone ran an assessing eye over their children; every stranger was a potential talent scout. (Josh Conroy had been spotted that way and now, despite the dyslexia, had a scholarship to one of the better universities; Cameron Hayden had been whisked from the locker room to children's television where he was granted his own weekly slot with Dr Woofs, a puppet dachshund who was an expert on pre-teen nutrition.) One day The Cougar took my forearm and examined it, nodding to himself, pursing his lips.

'You have the right bones,' he said, turning my arm over and looking underneath, pinching the skin to check for fat, assessing the width of my wrist against his own. It was simple physics, he said, addressing my mother now but still holding my forearm. 'The shorter the bones, the easier the lift.'

When he gave his card to my mother, she slipped it inside one of the secret inner pockets of her purse, as if it were a winning ticket.

Within a week The Cougar had started training me and my sessions with my mother had stopped. I had thought she might be sad about that, perhaps even refuse to step down, but on the contrary, she seemed glad of the time to herself, and whenever I saw her she was painting her nails, or trying her hair in a different style, or hanging new clothes in her walk-in wardrobe. Sometimes she sat in on our sessions, but even then she offered none of her usual advice or commands, and I had the feeling that although she was watching closely, it was not me she was observing.

If I wanted to make any real progress, The Cougar told me, I would need a name – something that made people think of strength, potency. Animals were good. Did I have any favourite animals?

'I like fish,' I said.

The Cougar laughed, but not in a friendly way.

'Um . . . I had an ant farm once. Ants are pretty strong.'

The Cougar ran his eyes over my pale arms and legs, my milky chest. 'The White Rhino,' he said. 'Very rare and very strong.'

I wasn't sure I wanted to be a rhino. They bathed in mud and had horns on their noses and little stumps for legs. I looked at my legs.

'We'll have some T-shirts made up for you,' said The Cougar. 'You can give them to your friends at school. Your mother and I will wear them too.'

I wanted to tell him that I couldn't think of anyone to give a T-shirt to at school. It was bad enough that I was so much bigger than everyone else; the boys were intimidated by my bulk, and the girls just thought I was a freak.

'I'll have another think about a name,' I said, and for the rest of that week I tried to find something that fitted. The problem was, all the ones I could come up with were already taken: The Bomb, T-Rex, The Hammer. Besides, my mother and I had been very successful up until then without using a nickname, a fake ID; why change what worked? I could not rid myself of the feeling that The Cougar wished to remake me entirely, and that I was expected to go along with the christening of this new being, this giving of a name to a child who was not quite human.

At school Mr Mayhew was teaching us about space. Even the word was exciting: it made me think of room, of vast quiet areas. I imagined drifting in the cool darkness, released from the pull of the earth, free of the undertow of competitions, training, public appearances, carb-to-protein ratios. He told us that although we might think space was far away, it was also very much a part of who we were. All things, he told us, including stars and even galaxies, expanded and contracted like a human heart. And all the calcium in our bones and all the iron in our blood were made billions of years ago, in huge blasts – he paused to use his special coloured chalk, drawing bursts of red and orange on the blackboard – called supernova explosions. The supernova, he said, was the death of a star, and could be bright enough to outshine an entire galaxy. After they had exploded, the biggest stars collapsed to form black holes, and inside these, gravity decayed. I had not thought such a place was possible; I had assumed that gravity, my oldest companion, my biggest rival, was indestructible. The thought that it could decay – that it could die, just as Mrs West next door had died – filled me with a joy that made me want to laugh out loud. I felt light, buoyant. Mr Mayhew was delighted by my many questions. He became quite animated as he spoke, his skinny hands flitting about him like white birds. I could have listened to him talk all day, until the sun disappeared and the stars came out to offer him their glimmering endorsement. I was disappointed that I had to go away for a few days the following morning, but I had been invited to open the new extension to the airport up country – dressed in a pilot's outfit – and The Cougar had said yes on my behalf.

When I came back to school I saw that the classroom ceiling was strung with tinfoil stars, lopsided papier-mâché planets, cellophane comets.

'You don't have to make a space sculpture,' Mr Mayhew said. 'It was just something we did for fun.'

The pipe-cleaner astronauts in their toilet-roll spaceships seemed hopelessly childish to me, and I felt as if I did not belong in this room, with these boys and girls who were all so much smaller than I was. I felt like a giant, as if I might break my desk if I leaned too heavily on it, or snap a pen in half if I tried to write. Whenever I returned from a trip away, I always felt as if more time had passed for me than for those I left behind; it was as if I aged more quickly than they did, so that the gap between me and my peers expanded, the boundaries hurtling away from each other. I thought about the sort of space sculpture I would have made. It would have covered the entire classroom ceiling, I decided, eclipsing the embarrassments made by my classmates. It would have been something to be proud of – something outside of myself, quite separate from my muscles, my flesh.

'I would have made a supernova,' I said, and Mr Mayhew smiled and said he didn't think we had enough cartridge paper for that.

That night I told my mother and The Cougar my name: I would be The Supernova.

'Isn't that a brand of instant coffee?' said my mother.

When I explained it to them, they agreed it had a certain something, although The Cougar said he still thought an animal would have been better.

He first mentioned the deadlift at breakfast one morning. I was surprised to see him there at the table, sipping his coffee and reading the newspaper like a father.

'I've been thinking,' he said, without looking up. 'You need something new, something to make people take notice.'

'I'm The Supernova,' I said. 'I'm the strongest boy in the world.' I pointed to one of my framed certificates.

He shook his head. 'Not enough any more. You're nearly 13; it's a dicey age.'

'The Cougar's right, darling,' said my mother. 'We need something new.' She stood at the stove, an egg slice glistening with oil in her hand. She never made fried eggs.

'The deadlift,' said The Cougar, and looked at me.

I wasn't sure what I was expected to say. We incorporated the lift into our training, of course, but it was unpopular these days. Some people thought it placed too much stress on the lumbar region, that the human body was not built to withstand such pressure. 'The deadlift,' I said.

He nodded. We needed to raise its profile, he said. Performed correctly, it was a beautiful movement; a combination of strength and dexterity that unlocked the great power held in the human form. The Cougar believed I could break the current junior world record by at least 40 pounds. 'You'll be ready to do it at the meet in October,' he said. 'You'll be a *hero*.' He gripped my shoulder, and already my mother was at his side, nodding, her eyes shining.

'It would be perfect for your profile, darling,' she said, and I opened my mouth to answer, but she was looking at The Cougar.

We began training that afternoon. I had been doing deadlifts for years, but The Cougar told me I must put aside all notions of the correct method. 'Imagine you are an infant just learning to walk,' he said. 'You don't know how your body works. You don't know where your centre of gravity lies.' He made me wear ballet slippers; they were black, not pink, but they were still ballet slippers. He said that the thin soles meant I was closer to the ground, reducing the distance I had to lift. It was simple physics.

That first day, he wouldn't even let me touch the weights; instead, we practised bending and rising, my hands in the right position but holding nothing. And yet, I could already sense the heaviness; I could feel the grip in my palms, the hot stretch in my arms and chest, the barbell a phantom limb. I bent and rose again and again. It was like praying.

'Remember to keep looking ahead, not down,' said The Cougar. 'If you look down, your neck is at the wrong angle in relation to your back. The angle is the thing.' He tapped at my hips. 'You are a lever. When you push your hips back to get more knee flexion, you move the fulcrum away from the bar.' He prodded my hip joint again. 'You're increasing the lever arm of the resistance, meaning . . . ?'

I frowned at my ballet slippers.

'Meaning that the bar becomes a lot heavier. Simple physics.'

At night I lay in my smooth, cool bed and wondered how much I could lift; if my strength had a limit. For if I could lift 500 pounds, then surely I could achieve 501; and if I could lift 501, then why not 502? The harder I trained, the stronger my muscles grew, so that the more I lifted, the more I could lift. If I thought about it like that, there was no end point: I could shoulder the world. I knew that if you plotted weightlifting records on a graph, the curve climbed like a gentle hill, the gradient decreasing over time until it almost levelled out. I pictured myself running towards the top of that hill, looking for an end point, a finish line, and never finding it.

Item by item, The Cougar's belongings started appearing at our house. A jacket left on the back of a chair; a pair of shoes at the front door, the laces still tied as if the wearer could not wait to come inside. There was an unfamiliar brand of soap in the main shower, too, a curling black hair stuck like a question mark to its surface.

I fully expected my weightlifting career to continue into adulthood; in those days I could think of nothing more satisfying than making a life of the competitions, filling my house with trophies until every wall glittered. I would be so wealthy that I would employ someone just to clean them. I would be on TV, perhaps even in the movies. I would sign so many autographs that my signature would be a squiggly line. After my sessions with The Cougar, though, I couldn't quite summon these images of my future. It was like trying to remember the face of someone you've met only a handful of times. Instead, I found myself looking back, thinking about all the titles I had won, the countries I had competed in. And then, when I tried to imagine the total amount of weight I had lifted in my life, I began to feel a pressure on my chest, as if something were weighing me down. I thought of the men who inhabited the fringes of the competitions, the one-time champions gone to seed, the paunchy coaches, the former stars.

I am often asked how I managed to lift so much; how I was able to bear such loads. Dr Sime, for example, wants to know how much I trained, and how frequently, and whether it was painful. My workouts did hurt sometimes, it's true, but it wasn't like the pain you feel if you trip and skin your knee, or if you fall on gravel and get tiny stones embedded in your palms. This was a feeling of taut expansion, a fatigue in the muscles and tendons and bones, the sense that I could not perform a single lift more. The trick was to distract myself – to deceive myself – and this was the technique The Cougar encouraged me to use. He told me I needed to find my own place, somewhere I could escape to in my mind. He went to the jungle, he said, and imagined himself creeping through ferns and vines, his powerful paws silent against the moss.

'Cougars don't live in the jungle,' I said. 'They live in the mountains. Mr Mayhew told us.'

The Cougar frowned for a moment. 'Well, Mr Mayhew isn't your trainer, is he?' he said.

I shook my head.

'Do you think Mr Mayhew could train you for the deadlift?'

I shook my head again.

'All right then. I want you to think of a place, somewhere that has a special meaning for you perhaps. It doesn't have to be somewhere you've visited, and it doesn't even have to be a real place.' He watched me, waiting for an answer. I was still thinking about cougars not living in the jungle, and thinking that if he was wrong about that then he could have been wrong about any number of things, perhaps even wrong about how much I could lift – but I said yes. My place was a beach, I told him, and he was satisfied with that – but in my mind I was remembering what Mr Mayhew had told us about black holes: that the gravity inside them was so strong, not even light could escape their pull. He described them as hellish locations, as dark demons that would swallow you whole as if you were nothing more than a tiny pill, but I did not think of them that way. In such a place, I decided, my lifting really would be a wonder; in such a place, the tiny increments that broke world records would lose all meaning. If I leapt into a black hole I would fall and fall, but I would fall slower and slower because, in the presence of such strong gravity, time would slow. Perhaps then I would stay the same age as my friends.

The black hole came to me then, growing and spreading like a balloon. It filled my skull, pushed itself down my arms and into my fingertips, gushed around my heart and into my stomach and legs. I didn't want to share this place with anyone; I didn't want The Cougar entering it, messing it up, leaving his things there. If weightlifting was a fight against gravity, if it was about overcoming the pull of the Earth, I would take myself away from Earth. I have never admitted this to anybody, not even to Dr Sime, but my mother made me walk too early; she made me pull myself up from the Earth before I was ready to leave it. The thing is – and I realised this cocooned in my black hole, which was not an evil place, not the hell Mr Mayhew had described – no matter how much I lifted, no matter how high I held the weights and no matter for how long, they always had to come down again. Gravity always wins; we all have to begin our downward journey as soon as we have reached the summit; we all become stooped, our bodies drooping lower and lower; we all return to the Earth.

*

In the sauna I melted into the smooth cedar walls. If I closed my eyes I could feel my lashes warm against my skin. Even the air was hot, so that you felt it entering your mouth and throat as a physical mass, like hot cotton wool. The October meet was in one week, and The Cougar was restless, pacing around me as I trained, his hands pushing at me, smoothing me, correcting my posture as if he were forming me from clay. I knew by then that he was sleeping at my house, in my mother's bed. I imagined his brown skin against hers, oiled and gleaming like a cockroach. Did he put his hands on her too, position her limbs and her back, rearrange her until the angles were correct?

The day of the meet I did not want to talk to anyone; I knew that if I let any chatter inside me it would displace the calm I had created, the quiet black. I think my mother was a little upset. Wordlessly she applied the fake tan, her strokes somehow mournful, as if they were a farewell.

There were a number of reporters hanging around, some with television cameras. They tried to catch my eye, ask me how I felt, but I would not let them in. My mother and The Cougar lingered backstage, wishing me luck, giving me useless last-minute advice. When my tan was dry The Cougar hugged me and told me to remember my simple physics. He felt smaller than he looked.

On stage I waited while the MC waved down the applause and asked for complete silence. Then I gripped the bar, wishing away the hall, the lights, the crowds of people, willing the black hole to replace it all. My mother and The Cougar were in the front row. They were holding hands, their fingers knotted together, their heads almost touching. I knew I had to keep looking ahead in order to perform the lift, but all I could see was The Cougar's hand holding my mother's. I could make out every knuckle hair, every half-moon nail. I began the movement, bending first and then rising, pushing up as if pushing myself through the Earth, and instead of looking ahead, I looked down.

Nobody knows if black holes are real; nobody has ever seen one. They are a theory only, as true or as false as heaven. I held the bar against my chest, elbows steady, forearms unflinching, fists framing my heart. I had the feeling that time was slowing, decaying; that I could have held the weight there forever; that to lower it – to complete the lift, to break the record – would be the end of something. I felt a twitch in my leg, and my thigh began to shake. I had no sooner steadied it than the other leg started to quiver, and the quivering moved up my body and sat in my chest, beating its wings like a trapped bird. My shoulders shook, my arms trembled, and I looked out to the audience and saw my mother's face, her mouth a round hole, her hands held out in front of her as if waiting to catch something as it fell. And I let go.

I don't recall the sound of the weights hitting the floor; I don't recall if they cracked the wooden boards or if the front row of spectators ducked. I only remember stepping back, back from the edge of the stage, back from the hot lights and from the falling weights, back and back until everything became small. The audience, my mother's face, the weights themselves: they all receded, until they looked so tiny to me that I could have cupped them in my hands, held them there inside that little black hole of my own making, and never let them go.

STUFF

Margaret Mahy

'Stuff!' said the old man, lying in bed with his eyes closed. 'Stuff! Me! And you!' He seemed to be mumbling into the ear of the woman bending over him, settling him down onto his back. She skilfully hooked her right arm under his sheeted legs, and his knees rose submissively – twin peaks, draped in the snow of a clean sheet. She hoisted him up just a little further.

'Stuff you too, dear,' said a second woman, who had been blotting him with toilet paper a moment earlier. She slid a bed pan away from him with practised skill.

'No!' said the other woman. 'He's been on about "stuff" a lot lately and I think he's just talking about . . . well, about *stuff!* Look at his hand there.'

The old man's slow fingers were pinching a fold of his bed cover, and rubbing it slowly backwards and forwards. His eyes opened.

'It's so astonishing,' he said, speaking with unexpected authority into the air between the two women. 'When I was a boy, you know, I planned to be Dan Dare, Pilot of the Future. I thought I would go out into space. Out!' He flung his arms wide. One of the women ducked. 'Hey! Careful, dear,' she exclaimed.

'If a body is at rest,' he said carefully, 'it will remain at rest forever unless acted on by an external force. Isaac Newton.'

'What? The falling apple man?' said the second woman. 'Well, he was right. And I'm an external force!' She shuffled him sideways. 'You'd lie here at rest forever if it wasn't for me and my partner over there acting on you.'

'Beyond the planets!' the old man cried. He chuckled, making a sound like a rusty lock turning. 'Funny that I should start dreaming about it again. Space, I mean! Out there! And all when I'm locked down here . . . nothing but leftover stuff these days!' His voice changed a little. 'Mind you, there's always the mystery of stuff itself . . . of stuff being what it is, I mean. Even the rubbish of the world is baffling, isn't it?'

'You lie there and think about it, dear,' said the first woman, rearranging his pillows. She looked across the bed at the second woman. 'Is he comfortable enough?'

'Nothing but stuff,' the old man persisted, frowning obstinately. 'But I'm still self-organising, mind you. And the material world isn't as definite as it seems to be. Solid matter dissolves away to be replaced by the laws of chance rather than rigid causality. I think that's what's happening to me.'

'He's fine,' said the second woman, and then put his bed pan on the trolley and they left him there, still sliding his hand over his bed cover as if it might have something to reveal.

'Astonishing,' he repeated, mumbling to himself now. 'Well, everything is. We all are. Made of.'

Then his right hand crept over to his left, to pinch up, once more, a fold of his own loose skin. He rubbed it softly, closing his eyes as he did so.

'It holds me in but, after all, it's only part of the . . . that other form of the . . .' He stopped struggling with a difficult thought and fell silent, though he was not asleep. His right-hand fingers kept working the pinch of skin backwards and forwards. 'Once I longed to go out to the planets . . . beyond the planets.' He laughed a strange, breathless laugh there in the pale room. 'I've been everything in ordinary life. I've been a building contractor. I've been a married man . . . a father . . . had three kids. I gardened . . . ran our local rugby team. Jogged to keep fit. Real life. Great! I wanted it all. But what about *true* life? One night, before all that real life took me over, I was walking home through the park. Hey! Remember? Remember, you in there? Remember stepping into that island of moonlight? Those trees deliberately bent their branches back so that I could stand underneath them, but still see the sky when I looked up – the sky and that strange, white, staring moon. And even though the moon up there was flooded with reflected light – it was a full moon – I thought I could make out the crater Plato. But of course I knew just where to look for it. I'd had that telescope back when I was kid, and I'd looked up and out, up and out, night after night after night. Anyhow, there in that patch of silver under the trees I began to dance – didn't plan to, just began to – this way, that way up and down, arms held high. Funny way to behave I suppose. Not as if blokes are supposed to dance in a patch of moonlight. My good luck that no one caught me at it. *Way out!* That's what I was thinking. *Way out!* And I held my arms up towards the moon. *Almost there*, I was thinking. Not *real* life but *true* life. Because, right then, I think

I was really close to finding the way out and up. Almost. Almost but not quite.'

His fingers crept up his arm pushing the loose sleeve of his pyjama jacket high. There on his upper arm he had a tattoo – a circle of leaves framed a small closed door. 'Way in' said the words on the door. He did not have to look at his arm to know exactly where that door was.

'Funny! I was only – what? – 17 back then, thinking all that *out, out, out* – nothing but *out*. And yet I had the words 'Way in' tattooed on me, and then a door tattooed around the words and a year later leaves tattooed around the door. Back then it seemed like a joke I was having with myself, but now it seems like I was really inviting myself in. Suppose, now I'm old, suppose that door opened for me at last . . . suppose . . .'

The door opened . . .

. . . and there he was, staring across a countryside, knowing he was the first person to set foot on that particular desert . . . so arid, so strange yet so utterly familiar. On it went, on and on, a strange, dry, empty landscape, broken only by blades of grey grass, until in the distance five peninsulas stretched out into some indistinct pale sea. The surface under his feet was unexpected. It was ridged; it was yellowish brown and freckled with patches of a darker brown – dead but not disintegrating. As he crossed it he saw that it was flaking a little. Every now and then a pale shadow would detach itself from the desert and float away, its edges crumpling as it drifted. Yet for all its inert fragility, for all its hints of vulnerability, the desert was closely linked into itself . . . was somehow filled with a calm assurance of its own necessity. Under the dry surface a river flowed – a blue river with blue tributaries feeding it.

The old man felt suddenly flooded with a feeling he had not felt for years . . . excitement and adventure underlined by the freedom to choose. 'Open sesame!' he muttered, confidently stepping forward (confidently walking the fingers of his right hand in small steps down his left arm). 'Way in!'

He reached the first river and walked out over it, for there was a secure membrane between him and that blue flow under his feet. He stopped mid-stream and stared down into it, only to find a featureless shape staring back at him . . . his own reflection, perhaps. He bowed down towards it and the shape seemed to bend obediently up towards him. 'Way in?' he said, asking a question as he extended a hand, and the shape stretched a hand up to meet his. In spite of the separating surface their fingers locked. He had imagined that he might somehow rescue that other possible self from the confinement of the narrow blue river, but he was the one who was drawn down. Within a dissolving moment the river was throbbing around him, sweeping him on.

Surprise! There had been no intimation that the sleek river would be such a crowded community. He had not expected that he would have to struggle so hard, bending left and right as if he were a man made out of ribbons, in order to find a place for himself as a thousand disc-shaped identities moved in on him . . . around him, over and below him. He paddled his hands, though his movements no longer seemed to be entirely his own. The swarm closed around him, urging him forward.

So there was only the one way to go. But there was excitement too, as he understood that he could not be excluded. Was he still lying in bed, propped up on pillows, with people around him secretly wondering why he bothered to stay alive when he had nothing to live for? He was inescapably in this river too, a point of energy, lively and self-aware, moving according to a necessity which he knew to be beyond his command but which seemed to be singularly his own. 'Way in,' he cried, not with any voice, just drawing on the essence of that old ambition which had always been part of him, and which was only now fulfilled. He laughed breathlessly. That 'Way out,' he had once cried as a young man, dancing and staring up past the trees to the moon, and then past the moon towards those stars asserting themselves in spite of the dominating silver light, had become an opposite cry. Yet, strangely, it was the same cry. 'Way in,' he shouted again, giving the words inscribed on his upper arm new necessity. Out there the planets spun, flicked by the sun's invisible finger, and beyond the planets lay space that seemed empty even though it was loaded with the debris of a mysterious first expansion, and was possibly connected with other dimensions as yet unobtainable. Now, squeezed by these disc-shaped entities, he felt he was advancing into a reflection of that outer mystery, contrary and yet connected. The river throbbed him onwards and he thought he recognised a rhythm he had danced to from his very beginning . . . a rhythm that echoed the hidden beat of time.

Then something enveloped him, wrapped him around in jelly. Not a jelly, he told himself, as a word from distant studies swam up from the depths of memory . . . a lipid. He was being devoured, perhaps, by one of those cells battling beside him along his inner river. But then, perhaps the only way to arrive at the centre of things was to submit – to let yourself be devoured. 'Way in!' he cried. 'Go on! Eat me!' Terrified but exultant.

He had expected to encounter mild resistance but he slid through that lipid boundary, only to find himself in an active electric city alive with messages, exchanges, commands, declarations, with continual building and demolition. It seemed he could actually hear it . . . the whisper, the thrum, the roar of its constant action, but what he was experiencing was something beyond hearing. This strange conurbation constantly structured and restructured itself around him – thrust him this way and that, built him into itself with the energy of its continuous self-creation, and then discarded him, tossing him off like a scrap of useless debris, only to bombard him and use him again a moment later. Long pods fed energy into the cell around him. *Mitochondria,* said that voice from his school days. *Almost like cells within the cell.* There was the feeling of a network closing in around him

'Once I wanted space,' he repeated obstinately, then shouted it aloud, struggling against the bombardment, knowing it was not yet time to surrender himself. It tore him apart. *'Hey!'* he cried, protesting. *'It's me. Me! Once I wanted to be up there . . . out . . . way out! And here I am reduced to going in . . . way in. There's no space here. No space! No universe! I'm being assaulted by – well, by myself I suppose.'* He had always imagined his cells to be a crowd of gentle entities, nudging one another but cooperating in a mighty construction . . . forming *him* as an unconsciously created work of art. And it might well be true, but it was only true in some ways, for this cell felt as if it was dancing a wild dance that was entirely its own, a dance he recognised since he had once danced a version of it himself, kicking up and out in that circle of moonlight. Electrical energy raged through him. Messages were being flashed backwards and forwards, but he had no time to read them . . . no great wish to read them either. They were to do with the motives and necessities of the cell. Swallowed in as he was, he felt himself to be an accidental by-product, for the cell was focusing only on itself, and yet he also knew he was the distant master. '*It's what I am*,' he thought, '*This crowded, jostling savage community is what I am.*'

As he thought this, the cell suddenly began to dismantle itself and he understood that, in a businesslike way, it was choosing to die. There was nothing ominous about this. Cells were always businesslike about their own termination. They had no fear when it came to self-surrender. But he was not as businesslike as the cell. Not yet. He must escape or die with it.

He battled on, writhing through protoplasm, to pass easily through the lipid wall of the fading cell and then, directly, into yet another cell and yet another zone of furious activity. Once again he, himself, the entity containing the cell, was also contained by it, reduced to a mere construction . . . an idea, perhaps, struggling to define itself. Beaten this way and that by the cell's necessities he struggled on, and then saw a shape ahead of him which seemed as if it might be a refuge of a kind. So he made for it as well as he could and found himself melting through its surface and becoming part of it.

And now he was in a strange world indeed – a world of spirals. Staring at them, he had the curious feeling he was seeing, not random patterns, but himself – a series of necessary commandments. Moving towards them he spun without being able to work out if he was spiralling up or down (but, after all, he had moved well beyond the up-and-downness of things), though he did feel even more intensely that connection with self. These curving, curious patterns felt as if they were part of the structure of the world, yet felt, at the same time, as if they belonged only to him . . . as if they *were* him. It was strange to encounter something so uniquely personal which simultaneously seemed so universal – something so singularly private that also felt abstract.

But he was a committed man. 'Further in! Way in!' he told himself sternly. Then, leaping to identify with a particle that was moving past him, he did indeed move further in.

'There must be a way of understanding this,' he thought. 'Hold still just for a moment. Hold still!' he commanded. But nothing held still. In fact, if anything, there was more movement, for he became aware of tiny structures shifting around him. 'Molecules to let,' he thought. 'That's what they must be. And the molecule is made of atoms. I am almost there.' He had moved from one reference frame to another and was now in a world of possibility. 'Is there any reality in the microworld? Does this really exist? Because here I am, wondering, and if I wonder I must be real in some way. But this place is governed by laws of chance rather than causality. I am setting myself free from cause.'

For, once again, the world had changed around him . . . if it was a world, that is. It was busy, but no longer frantic. 'Atom' means indivisible. Close! So close. Not to the longed-for contemplation of space, but to the opposite of space. Yes. There it was.

In his head he saw a picture of that tattooed door. For him the atom would not be indivisible. Way in. So now, as a point of curiosity, he went even further in.

And suddenly everything around him expanded, becoming not just an open field, not a forest or an ocean, but an endless cosmos. There in front of him was its centre of power. He had surprised an inner universe that seemed to reflect something of the universe out there, that universe which real life had denied him when he was young and active. Nevertheless, he had won his way, at last, to huge space and silence, though there were paradoxes around him, for somehow that silence came in on him as a sort of sound . . . a sound he had no name for. He was obliged, after all, to describe it to himself as music, though even as he named it he knew that neither music nor silence was the right word for what he was hearing . . . and hearing was not the right word for the connection he was making. Particles have frequencies, he told himself. Perhaps they *are* notes in an abstract musical system. Not only that, suddenly he was no longer an alien observing fragment but was spread out through it all. Could you go further in when things were somehow *there* but being *not there* was just as significant?

I am a particle, he thought, but knew as he thought this that he was not just a particle, for a particle occupies a specific point. However, he had extension – extension throughout this universe and perhaps into other dimensions. He had become a wave . . . a wave of possibility . . . and, moving like a wave outwards, there and everywhere simultaneously, he found himself expressed in a form that seemed, by now, beyond identity. Yet he had identity. He was a singular excitation of the field in which he found himself, moving into some topological state he could feel but could not name. But perhaps he was beyond naming. Names pegged you to a time and a place and inexorable identity.

'Space,' he said, in wonder and recognition. 'I'm not *in* space. I *am* space. It's what I've always been.' And then he thought, 'It's restful here, at the heart of the heart, being stuff of stuff, so perhaps I'll remind myself of the song. Perhaps I'll join in. And perhaps I'll take a rest, even as I sing and sigh and spin . . . a true space traveller at last.'

WWW.
AND OTHER POEMS

VINCENT O'SULLIVAN

Part I

Dr Newton's Sock

Hommage à Science

The idea of it takes off, orbits,
trajectories predicted as bare fact
recedes. All becomes mathematics
doing as it does, the equation
that masters, replaces, such things
as described.
        'The thing in itself'
as we used to say, poets who preened
as savants in our coloured hoods,
in our wooden masks, in a line
drawn in sand on whose other side
the prosaic spaces where science cavorts,
intent on defining what it says it sees,
what the mind goes putting in place
of already there, a world figured
in ciphers. How can we know, we
feathery dancers,
              we rhythmic stompers?
Erase the line in the sand, say
'The floor too is yours, we step in each
other's shadows, light and dark at play.'
And turn to our separate mirrors for what,
whatever the ending, starts the same.

Every dance is an excuse me

To declare what is anciently there
becomes newest excitement, as the Parthenon
say revealed yesterday at noon,
a quartet uncovered in a cupboard
we thought was jam, a duck flung
from a *chapeau,* its quark music
of a kind. All things circle and flair
and relate even when they don't.
It's a dance with curious partners
at the *Palais des Sciences.* Their names
may be different tomorrow. So may yours.

Science, to begin with

I like the stories, although the stories
are not what it's about. I like
to know Lavoisier lost his head
to a meaner scientist, that Cavendish
squeaked and ran from human dimensions.
Or Rutherford as a boy when his mother
tells him, through a storm, what makes
lightning strike, he answers politely,
'No, no it doesn't, mum.'
                                    But that
is like liking the wrapping wrapped around
the gift, the gift as much in the dark
as the famous cat, that mascot in theory
at least of every lab – 'Things can go
either way, although only one.'
                                    Reality's
a tall order the further we descend,
though starting at either end almost hits
a rhyme – 'Adam,' we like to urge,
take all this in.' Atom, as we say
as well, let's begin from here?

Elementary love poem

It is weird enough, to come back
from the minute places, from the words
we are used to even in Third Form,
proton that packs at the centre, electron
careering in unapparent space.
                              All that,
leaving all that, to return to the big-
time jangle of where *we* fit in,
to what we're presumably made for (so
to speak), yet nostalgia too, how
beautifully things arrange in those micro-
vistas, nano-expanses: the Russian doll
syndrome of this into smaller thised
to smaller yet: the bits of us, love
– I have to tell you this – the bits
that will truly last. Unearthly's right!

Madame Curie among the Menfolk

The facts remain before we know the facts.
She is saying, to make them smile, that
masculine crew, 'I unhem creation a little,
to work out the stitch.' Then giving
the stitch its name, that's the easy part,
labels that keep in fashion once applied.
'Mind bringing it together, that's sewing too.'
And another truth emerging in the clock of her bones,
Radiation hardly a name to be handled
lightly, although that's her shadowy story
only time will tell. And a whisper
from old-time Europe, 'You think that a bargain?'
'If it has to be that, then yes,' she tells Dr Faust.
Her lab books sealed in lead. Holding something back.

Dr Newton's Sock

A deep craving against chaos, the body
itself demanding that it pay attention,
that it know its place, the transports
of 'to be' among what it describes,
and so understands, a matter of gravity
call it, as the mind revolves. . . .

Newton for several hours on the edge of
his bed, drawing on his sock; his sock
and the cosmos kindred. 'Law,' he divines,
'makes equals of us all.' So the penny
drops!
          As he stands there, Dr Newton,
the full-dressed man, mind saying
*This is as it is, which means as should
be:* God's finger indeed, as he's proving
the finger not required.
                              As trivia swirls
about him, tides of mundane trash,
housekeepers irate at delays that hold
breakfast up, the world and thought-of-the-world
descant each to each. 'Where have you been,
Dr Newton, in your unmatched socks?'
'Pretty much all over, madame,' a mere fact
he lets fall.

'Into the depths . . .'

There is a world we come to where touch
lets us down. The professor of close readings takes
its measure, writes E for so much of what is,
m for so much else. m he says is E
waiting to occur. He knows so, precisely.

Were Dante still tour-guiding he might work for
Physics, advise 'Watch every step as I lead
you on. Calibrate my quill-tip carefully,
observe my row of noughts, their descending
tunnel.' And the Southern Cross, he instructs,

still there Easter Morning, although one
might chance one's arm, mention various gases,
offer a reasonable hunch for when stars
run down. Which may not surprise him
a jot, that laurelled Maestro, 'There is

room for God knows what in the unfolding
Rose,' cute as a cat watching his favourite
goldfish. Measure it through and through we still
call it fish; track its riddling neutrinos,
it will taste the same. 'Same' being a net

so wide when it hauls Poseidon
by luck from the silt of centuries, its components
the physicist says are exactly these, its humanity
perfection of another kind. And the great bronze arms
stretched back, and forward, and blessingly over both.

A Simple Man's Quartet

In a much simplified way, these poems relate to an experiment in particle physics. Its basic fact is that antimatter atoms mirror their mundane counterparts in every way, except for having the opposite charge. Antimatter particles meeting with matter particles release enormous energy.

In the experiment, anti-hydrogen atoms – or H-bar atoms as they are also called – are formed in a 'trap' that contains both protons and antiprotons. These atoms escape from the trap and hit the walls of the container, where they are annihilated. As this occurs, two gamma rays are produced, and these in turn produce high-energy particles called pions. When these are detected they confirm what has just taken place.

The next consideration is how this might be used as propulsion. NASA provided the figures that 42 mg of antiprotons possess energy equal to the 750,000 kg of fuel in a space shuttle's external tank. So far as the experiment is concerned, the challenge is how to convert to thrust the energy from annihilated anti-hydrogen atoms. In theory, anti-matter pellets could power a sail of uranium-coated carbon, which would provide the fission necessary for propulsion. At the moment however this is purely speculative, and the manageable energy is minute.

(This sketch of the process is drawn from Graham P. Collins, 'Making cold anti-matter', *Scientific American*, June 2005, pp.57–63.)

A Simple Man's Quartet on H-bar Propulsion

i

Rap for the anti-hydrogen atom

You trap your particles you let them spin
they nest down together (there are millions let in)
there's a chance that comes off when the positron
runs the same trajectory as the antiproton
they begin to orbit next thing they're the same
the anti-hydrogen atom's right there in the game
(though the scientists don't like it they don't like 'just as'
but this is how it happens, and then you has jazz!)
Anti-hydrogen's crazy like it's hitting the wall.
Trouble is soon as 'is' means it isn't at all.

ii

To put it more sedately

Or rather, to put it more sedately,
this Johnny, this atom-come-lately,
disintegrates on impact, as its tendency,
two gamma rays emitted at characteristic energy
(for the expert among you, S11 ke V).
These what you might call consequential fly-ons
create their own particles the trade calls pions.
That your anti-hydrogen atom apparently doesn't
exist any longer doesn't mean it wasn't
the cause of what now is inevitably detected.
The evidence lies in how impact deflected
when the atom clashed with electrons at the wall.
That calling-card signed 'pion' explains it all.

iii

A small sermon on actual achievement

The kingdom of science, brethren,
not unlike the kingdom of Hollywood,
or even the kingdom of God,
is so nicely contrived of hope
and the glittering cities of what may be,
that to miss out on tomorrow
seems we're cheated today.

The fabrications of theory let's say
erect such exquisite promise
who cannot, who will not believe?
The sheer reasonableness of whatever
may never exist, the seductions
of 'proven on paper' already
the crown we wear.

That antiprotons then, seeing
they're on our mind, 42mgs
we'll imagine, energy to burn –
which equals, imagine again, the bulk
of space-shuttle storage,
75 thousand kgs, to take
a round number.

So how does that lift you,
believers? How does that propel you?
Nicely, we're likely to say,
Lord, we're on our way! Antimatter
in H-bar pellets, just how
far is far? If not, repeat, if not
for the drag of fact –

like intention converted to merit
on the way to salvation,
as tapping in front of a mirror

isn't quite Fred Astaire,
energy from working it out's
a long way from thrust. Yet we'll draw
a picture, brethren,

propose let's say propulsion, our
pellets rigged, our fission
triggered on our clever sails.
And the fact it won't get us lift-off?
Well, who's leaving before lunch?
Sufficient for the day, we intone,
is the day's antiproton.

iv

So to start again

What we labour to say, until compelled
        to say it,
is 'Trust how the mind and the world converse
        between them.'
Pursue where the evidence leads, trimming
        disbelief,
the neutron say jigging its orbit, too small
        to conceive,
yet equation riding its spin, mind working
        its way.
To be at home in the world, as we also say.
        To confront
the unexpected with 'We're here
        as well –'
is query stripped to its bones, the point of
        it all,
to say *This is so* as the mind impresses
        design,
when the first phrase is said as it were whether
        God's or ours,
'Let there be light' thought through to
        'The facts are these.'

Not Leaping but Waving

A few weeks before Christmas 1925 Erwin Schrödinger, Professor of Physics at the University of Zurich, took two and a half weeks vacation in a Swiss alpine town, accompanied by an old Viennese girlfriend (whose identity remains a mystery) and two pearls. He placed a pearl in each ear to screen out distractions, and set to work on the structure of matter. His 'wave equation' was published a few weeks later, displacing the assumption of quantum mechanics that electrons leap from one fixed orbit to another, without resting between these states. What was once a discrete particle is now to be thought of as a continuous wave.

Professor Doktor Schrödinger, overdressed, concerned
that the world concede him its due, which it never does,
is very particular, *nein?*
                    He leaves Zurich however
en vacance, he takes to a mountain village with a lady
in tow, a mistress from Vienna, from years back, *ach,* but years.

He carries two pearls, and intends to rut, to think,
both very seriously, for two weeks, and more. The Unnamed Lady
who offered her undulations, the alpine scenery
at every turn, worked wonders as we know
for the Advance of Mind: woke him one morning to soluble
brilliance, the wave of ziggy mountains, himself a mere dot.

The story is so simple it brings tears to the eyes:
the lady out walking, say, in the boudoir resting,
pouring black *Kaffee* while his pencil tapped: a pearl
in each ear blocking reason from distraction: the nature
of matter revising in mountainous, pure air.
Evolution clicks forward. Nuzzles Aryan bust.

There are two pearls on his desk as the world again
pours in. 'I have been on the crest, my dear,'
while the curtain bellies fluid, a moment as ever
on the move. '*Mein exquisites Atom,*' stalling
above her like surf. She closes her eyes to the drift
of precisely this:

'the particle as we thought it
is indeed a wave: which was leaping electrons, *Liebling,*
as they used to say: which is sequence flowing
exactly, say that from now on.' The biographer
detecting sex as subsidiary text. The equation
however standing 'lovely as man has devised'.

In the train down from the mountain the afternoon purred
like a cat. The Doktor's hand and the lady's ravish
existence to quite *this:* the trajectories of desire
demand their aural pearls. In his pocket the careful
pages shift the future about. 'A force even sweeter
than fact, its tide, its beaching, drenches us *here* as *there.*'

The lady has skimmed off as biographer's puzzle, her Doktor
– apart from that fortnight – a wavy male blur.
Guten Tag for all that, peaks salubrious at the window,
the paper completed, the woman dressing, the Professor's tie
tied. The quotidian resuming, the truth floats for all that.
Whichever hand opened, a pearl brought back.

Part II

www.

The Old Story

Arachne was what they called her. It meant no more
than the name of a girl who sat spinning,
though you might think a goddess sat beside her,
saying 'vary the colour here' or 'triple
strands are what you need.' You'd hardly look
at the cloths she perfected without thinking
how rich the world was, if this was a part.
                              How Zeus's wife
raged against her, a mortal coming at this!
Next thing Arachne, deft enough it was said
to detain sunset by a thread, to warp
the moon as icy witness – next thing her limbs
crumple, her wrists sprout fur, her head
dwindles to a sequin, 'human' far off
as a word in a language she had never learned.

Yet she's spinning all right, the once splendid Arachne!
Spinning the dawn's disc, the morning's cup,
between trees, across doorways, at the back
of wardrobes, above pillows so children scream
and the house wakes up. Compulsive
as the curse compels, she is spider, merely,
for those forgetting what she was.
                              'Spin,' we say,
'spin', as we might 'good girl' to a child
to keep her from tears, as we say 'that's clever'
of simple things – to be herself, which is
this: to glint from her baleful corner,
to weave divine malice into strength, into
glitter, into rare design: to say 'thread', that
solitary, surviving word. As though saying 'life'.

125

In the Beginning, before the word

Note: Proteins, the basic materials of spider silks, are composed of different types of structural units. These three 'pre-poems' illustrate the types known as amorphous, helical fibre axis, and β-parallel fibre axis.

1

This is neither a web nor a poem, but as a way to begin describing a thread, it is not impossible that the line becomes part of a poem; for poetry begins in fact, fact moves to knowledge, the delight of poetry is also the knowledge of fact. The first word is protein, to spin 'In the Beginning' ...

2

The amorphous may become helical,
The amino acids arrange just so.
A spider does not comprehend it
Nor the writer spinning his line
As the scientist instructs how Arachne
Proceeds with hers. Trust the spider.
Believe the scientist. Writing walks on air.

3

The β-parallel form fibre
\   \   \   \
hangs more like a Venetian blind
/   /   /   /
than this can actually bring home;
\   \   \   \
'plaited-sheet proteins', think of them
/   /   /   /
held in place by the bands that run
\   \   \   \
between the slats. This isn't of course
/   /   /   /
science and it's hardly poetry either
\   \   \   \
but one has to begin somewhere
/   /   /   /
if you're hoping to end with Silk,
\   \   \   \
let's call it, spangling to more
/   /   /   /
than it ever began as, as any reputable
\   \   \   \
spider or poet or scientist surely knows?

Read Instructions Carefully

1. In order for the web finally to dazzle
   first compress liquid protein though nozzle.

2. The pressure gradient on the way down
   applies tension as the silk is drawn.

3. The precise shape of the producing gland
   determines the angle of the molecules to hand.

4. Orient and shear the molecular chain.
   Polymerise the fibre again and again.

5. Control of the shear rate will determine
   the thread's surface and breadth. Now spin.

The point of it all

i
to angle precisely
   to handle the wind
      to make space home;

the spider as time's
   survivor, geared
      to its last atom

to string things
   along; *life,* carried
      smoothly on.

ii
to dissipate
   to slow down
      to freeze once and for

all; to harness the big
   fly the lordly buzz
      the small matter

of parcelled neatly,
   death, snapped
      in silken cuffs.

Laying the line for Ms Arachne

A figure as they say in myth who is sinister
or benign, denoting the moon as friend or stranger,
guardian or terror at the gates –

even our brain enmeshed in her metaphor's
hood, arachnoid* has us covered, hammock
and shroud together, closer to ourselves

than we'll ever come. And silk, as we smoothly
call it, the amino acids, the protein strands,
the molecular chains, 'here's beauty'

so we say, its death-sail raised,
its prismatic wheeling in sunlight, its spangled
lure, its signatures of self mathematically

set. And prediction micro-fibred to its finest
line, the formulae, the structure,
the castle of 'so it is' impregnably so,

the unalterable tilt of how its atoms move,
the flow and tug and spin that perfects the yarn,
the equation's cradling fact: *in all cases, thus.*

* A membrane enveloping the brain and spinal cord.

Light relief

From the human point of view
   – Which means everything said
Of configuration and structure
   From A to Z –

The 'cunning' as the layman calls it
   And science the protein chain
Permits a fibre to stretch
   And retract again.

The glow-worm may be cited
   As the fibre axis kind,
Unfolding then refolding
   As does a venetian blind,

Its pliable give and take
   A constant silky fact,
Allowing the strands to absorb
   The prey's aerial impact –

A bright engrossing feature
   To those in the know.
Without it, *arachnocampa*
   Luminosa's hardly worth the show.

Nice work for the Bristletail *(Thysanura)*

Evolution as expected is a strand that moves,
a matter of doing twice what surprises once,
to ask of an organ *this* as well as *that,*
its dying out or taking another chance.

Sex's particular fix was question and answer,
the female struggled, she'd rather be off and gone.
The male secretions were threaded to hold her down
(a case of the chauvinist taking a line of his own).

Collateral glands were the story, as this one shows.
Silks would get finer than this from a rough beginning.
Fibres will be delivered by other means.
Yet the oldest of urges sets a new world spinning.

Matters for future attention

Look carefully at the spider.
It looks back, I presume, at something
it knows is living, is not itself,
is too big to haul in,
would be better elsewhere.

Its components, its structures,
whirr and coalesce pretty much as mine.
It as much as I has 'genetic hot-spots',
so much for both may indeed
go wrong, on the way to right.

I read how its marvellous gift
evolved through 'inter-gland competition'
– Beauty, never forget, always has another
name. What's humdrum, compulsive
for spiders, makes me shiver

at dawn, the jewelled wet netting
across the garage door. Is there something
I offer it as remote from me?
One thing at least in common: 'cryptic'
and 'display', our categories of deceit.

'Just forget oneself!' science
instructs me, 'there's space for minimal
ego in the atom's spin. You may
watch the spider *qua* spider. Leave
it at that. Let the spider win.'

DEAD OF NIGHT

Witi Ihimaera

with Howard Carmichael
and David Wiltshire

1.

The table is set for six. The host, Captain Walter Craig, has invited the guests for pre-dinner drinks at 7.30. Time, of course, has long lost all real meaning, but on board ship a 24-hour day is still observed; it continues to locate, structure and define, calculate the days and measure the distances.

The ship is called the *Endeavour*. It is named after the vessel which set sail for Tahiti under Captain James Cook in 1769 to observe the transit of Venus across the sun. Cook also had secret instructions from the British Admiralty to search for the great southern continent which some 18th-century scientists claimed must balance the great land masses of the northern hemisphere; instead, he discovered New Zealand.

The purser is standing beside the table. He steps to one side so that Captain Craig is able to approve the setting. For this, the last dinner of the voyage, the purser has laid out the finest silver and, in the middle, a display of beautiful crystallised white orchids; the wine glasses are antique Waterford. Written in ornate script on place cards are the names of the guests:

Mrs Joan Cortland
Professor Van Straaten
Doctor Eliot Foley
Monsignor Maxwell Frère
Miss Sally O'Hara

Mrs Cortland will sit on the Captain's left. She will enjoy that, as it will give her a de facto role as hostess. Then, clockwise, the purser has seated Professor Van Straaten, Dr Foley and Monsignor Frère; he has carefully ensured that the Professor is not seated next to the Monsignor. Miss O'Hara is seated at the Captain's right.

'As usual, perfection,' Captain Craig says to the purser, noting that he has the ladies on either side of him.

'Thank you, sir.' The purser gives a slight nod. He is amused to see the Captain rearrange one of the dessert forks.

'What's on the menu this evening?' Captain Craig asks.

'The chef is offering a choice of entrée: either twice-roasted quail or a Balinese warm salad of lime, coconut, beans, spinach and shallots. The mains tonight are either seven-spice duck done with Singapore flavours, or steamed fish presented à la Polynesia and served with eggplants in beautiful lacquered purple coats.'

'And dessert?'

'Dessert will be either lychee jelly with green tea panna cotta, or Pacific fruits including mango, guava and banana accompanied by lime ice cream.'

'What about wine?'

'May I suggest the Tohu 2004 Gisborne Reserve Chardonnay? It displays peach and stone-fruit characteristics and has a light touch of oak which will complement the Asian-Pacific theme of the menu.'

'We have Tohu 2004 in our ship's cellar? Professor Van Straaten will be impressed.' The Professor, in addition to being a Nobel Prize-winning physicist, is a connoisseur of fine wines. 'Are our guests awake?'

'Yes, sir,' the purser replies. 'Monsignor Frère complained of queasiness from the last, somewhat rough passage, and Dr Foley joked that he has still to regain his sea legs, but all the guests are dressing for dinner. Mrs Cortland has advised that she will make a late entrance.' Mrs Cortland has both the beautiful woman's flair for making the most of the moment and the expectation that it is a woman's prerogative to be the last to be seated.

2.

As usual, Monsignor Frère is first to arrive. Resplendent in his ecclesiastical robes, he shows no sign of illness. 'I find, Captain Craig,' he says as they shake hands, 'that prayer aids recovery much more quickly than an aspirin.' He notices the dinner music and winces. 'We are beginning with Miss O'Hara's choice, are we? Ah well, the young and their tastes!'

At that moment Miss O'Hara herself arrives, gallantly escorted by Dr Foley. This evening she has dressed in a unisex retropunk style: a clash of vivid greens and purples over basic black that was fashionable at the turn of the millennium. As soon as she hears the music she begins to bop and gyrate.

'Hi, Captain,' she calls. 'Do you know this music? It's from an album made by Sonic Youth in 2005. Aren't they fantastic?'

Miss O'Hara had been the only one of the passengers whose well-being and comfort had worried Captain Craig. After all, she was an actress in popular cinema, and much younger than everyone else – but right from the outset she has been a joy and a surprise. Both she and Mrs Cortland have proved ideal antidotes to the lugubrious academic male company. 'Seldom has an underground band been so archetypally NYC in their music,' Miss O'Hara explains to the Monsignor. 'Don't you think they're sublime?'

Mesmerising instrumental passages, hazy indie ballads, devastating blasts of noise and free-form textural riffs build to a climax. It's all so New York City: dirty, sprawling, complex, harsh, self-obsessed, art-damaged, beautiful and cool as hell.

'Yes,' Monsignor Frère answers. 'Sublime.'

Next to join the dinner party is Professor Van Straaten, still fussing and apologising over his appearance. He has mislaid his dark suit and, instead, presents himself in tails.

'What, no top hat?' Monsignor Frère jests.

Miss O'Hara and Dr Foley exchange glances: it is more likely that the Professor has been put out by the Monsignor's sartorial elegance at previous dinners and has dressed formally to provide fair competition.

True to form, Mrs Cortland arrives last, just in time to button the Professor's wing collar and help him with his bow tie. 'You need a wife,' she says.

'Are you offering?' he asks.

Mrs Cortland appraises him with an ironic smile. 'You dear man,' she says noncommittally. Through her marriage to nanotech magnate and condensed-matter physicist Peter Cortland she became one of the richest people in the world. Now a widow, and still beautiful, she is accustomed to offers of marriage, though primarily from other billionaires more interested in a corporate merger. She turns to Miss O'Hara. 'I am so glad we don't clash.' She is wearing an evening dress of shimmering blue which contrasts stunningly with her red hair. The two women look like mother and daughter.

*

A waiter arrives with flutes of Laurent-Perrier on a shining silver platter, and Captain Craig orders the radiation shields across the windows lowered.

'Hemi, let's look at the view,' he says.

The shields slide down to reveal the *Endeavour*'s beautiful light-wings, like sails billowing. But it is not wind that is propelling this vessel. The sails are catching subatomic particles generated by the ultra-energetic jets spat out of black holes, particularly supermassive Kerr black holes, in the centres of active galaxies.

The ship is an Artificial Intelligence of the latest design. It wears its name proudly. In the days of 18th-century sail, Captain Cook had sailed its namesake from Plymouth knowing that there were a certain number of days to the Bay of Biscay. From there the distance and time would have been calculated by sextant and compass, and the days of westerly sail that it would take before they rounded Cape Horn. Again, distance would have been calculated to the South Pacific and time measured by days of sail.

Although 400 years separate Cook's voyage and Captain Craig's, nothing has really changed. Computer may have replaced compass, and distance and time are defined in light-years and parsecs, but bearings are still taken and distance measured by setting the destination: from Earth via Matariki to the Magellanic clouds, 50 kiloparsecs at the far edge of Earth's galactic disc; then across the Milky Way, 130 kiloparsecs to the Great Square of Pegasus and the Palomar 13 global cluster buried beneath it. From there the ship has jumped 670 kiloparsecs to the Magellanic stream and the Andromeda Spiral Galaxy.

Whenever Captain Craig wishes to convey instructions to *Endeavour* he does so through Hemi, its avatar. The Captain has a communication implant in his brain and, courtesy of nanotechnological advances, needs only to issue his orders from wherever he is standing for the Hypertime Engineering Matrix Intelligence to reply.

'Thank you, Hemi,' the Captain says now. 'Please maintain current speed and smoothness during dinner.'

'Yes, Captain,' Hemi answers. 'May I say that the ladies are looking especially lovely this evening?'

Captain Craig channels the avatar's voice into the ship's sound system so that all the guests can hear Hemi's compliment.

'You flatterer,' Mrs Cortland responds.

'Enjoy your dinner,' Hemi says.

Captain Craig and his guests resume looking at the view. The *Endeavour* is travelling so fast that it leaves a phosphorescent trail through the space-time continuum. The view is corrected by the ship's computers for relativistic distortions. A slide show of the passing galaxy clusters flashes up in a never-ending kaleidoscope of forms.

'I wouldn't be surprised to see dolphins,' Mrs Cortland laughs. 'Moments of beauty, like this, make me proud that ships are called she. Hemi won't mind, will he?' Then, uncharacteristically, she shivers. 'We're almost there, aren't we?'

'Yes,' Captain Craig answers.

'Ah!' Monsignor Frère clasps his hands together, pleased.

As his choice of dinner music, he has chosen Johann Sebastian Bach's cantata 'Ich habe genug'. When the great baritone Hans Hotter begins to sing the aria, he closes his eyes with pleasure.

'Now there's sublimity,' he says to Miss O'Hara. 'Bach wrote this cantata in 1727 for the Feast of the Blessed Virgin Mary. Hotter recorded it in 1950 and his interpretation attains utter perfection. Listen to the peculiarly profound utterance, the warmth and sincerity and the hushed, awestruck *mezza voce* above the solo oboe and murmuring strings.'

'The good Monsignor is already transported to Heaven,' Professor Van Straaten whispers in a voice which, unfortunately, carries.

Quickly, the Captain turns to all. 'Ladies and gentlemen,' he says, 'shall we go in to dinner? Mrs Cortland, may I escort you to your chair?'

It is enough! My consolation is only that
My Saviour, Jesus, may be mine,
So shall I escape from all the sorrow
That enslaves me in this life

3.

The Asian-Pacific theme of the dinner is a total success. Professor Van Straaten has gone into raptures about the 2004 Tohu Reserve.

'I'm not the only one to be transported to Heaven,' Monsignor Frère mutters to Dr Foley. But he, also, is enjoying the quality of the wine. The subtle flavours reach his taste buds and delicately rinse away the last of the chemical life-support fluids that sustained him, and the others, in deep sleep. 'How long have we been in suspended animation this time?'

'Better not to know,' Dr Foley advises him. 'Could your mind accept it?'

Endeavour possesses a photon epsilon matter-antimatter drive which uses the Penrose process to mine the rotational energy of the Kerr black holes, the energy caught by the ship's light-wings. Boosted by this extraordinary power source, *Endeavour* is able to travel arbitrarily close to the speed of light.

The Monsignor, however, will not be diverted. 'Tell me,' he insists.

'We are now cruising at 99.99999999999999999996 per cent of the speed of light,' Dr Foley answers. 'Therefore, if one uses human reckoning, by relativistic time dilation one hour by ship time is 126 million light-years.'

Monsignor Frère blanches. Prior to embarking on this voyage, he was the principal cosmologist at the Vatican Observatory. He understands something of Dr Foley's information, but the magnitude of the mathematics still staggers him. 'We've travelled that amount of time, every hour on ship, regardless of whether or not I am asleep?'

Dr Foley nods. 'More or less. Every time we've stopped for sightseeing, the slow-downs have been a really complicated series of manoeuvres; the time dilation becomes much, much less. Without those sleeps, before and after the stopovers, you and I would have been history long ago, so to speak. But they add millions of years to the time that Hemi has experienced.'

And the elapsed universal cosmic time since they left Earth? By now it would be several hundred billion years. Even Dr Foley's mind has gone into arrest as he contemplates this fact.

Miss O'Hara presses Dr Foley's arm and brings him back to the present. Professor Van Straaten is holding forth on the art of wine-tasting and requires everyone's attention. He asks the wine waiter to bring him a medium-sized wine glass with a rim that is narrower than its base. 'This is the best shape to hold the wine's aroma,' he explains. He instructs the waiter to fill it to just below the widest part of the glass and tilts it slightly against a white background to look at the true colour of the wine. 'It should be clear and bright and never cloudy.'

Everyone watches, amused at the theatricality of the Professor's presentation. He sniffs the wine and swirls it gently in the glass to maximise its interaction with the air. 'One must think about the differences, if any, before and after swirling,' he continues, taking a couple more long, controlled sniffs. 'One must ask oneself, "What is my first impression? Is the wine dominated by fruit smells or something else? Is it easy to identify or is it a mixture of things that make it more complex?" Wine should never smell vinegary or mouldy, although earthy can be okay.' Next, he tastes the wine by taking a small mouthful, moving it across his mouth and breathing a little air across it to intensify the flavours. He nods his approval and gestures to the waiter that he may fill the glasses of the others. 'Pay attention not just to the initial taste,' he says, 'but to the flavours that linger after you have swallowed – the *finish*.' He raises his glass in a toast. 'To truth,' he begins, 'and beauty,' he ends, saluting Mrs Cortland and Miss O'Hara.

To while away the time between one near-to-light-speed boost and the next, Captain Craig and his passengers have taken to playing a game during dinner. It was Mrs Cortland's idea, and they first played it when *Endeavour* reached the perimeter of the Local Group, 1000 kiloparsecs out. From there they had made a quick jump to the Triangulum/M33 and onward through the multi-faceted spectacle of the Virgo and Coma clusters. Dancing along the black holes in the centres of the galaxies of the Great Wall, and gaining vast energies at each encounter, the ship had shot off towards the Hubble Ultra Deep Field that marked the outer perimeter of the known universe. Once expansion of space was allowed for, it was a journey of 51 billion light-years.

The game requires that each guest tell the others what they think has been the most transforming event in the history of science, and, in particular, cosmological science.

'I think it's your turn to begin, Eliot,' Mrs Cortland prompts. 'As a prize-winning historian, your offerings have always been fascinating, and I am sure you have left the best till last.'

'In fact,' Dr Foley replies, 'I have an overture to begin with. Hemi, maestro, please!'

Dr Foley has a singular sense of humour. The first bars of Richard Strauss's tone poem 'Thus Spake Zarathustra', featured in the opening moments of Stanley Kubrick's *2001: A Space Odyssey,* roll grandiloquently through Hemi's sound system. Everybody starts to laugh. They laugh even louder as Dr Foley turns to Miss O'Hara and says, 'The next part is for you, darling,' and the Strauss is replaced with 'Good Vibrations' by the Beach Boys.

'Oh, Eliot!' Miss O'Hara blushes. Although Dr Foley is 10 years older than she is, he is actually the youngest of the men on board and they have fallen in love. In the last deep sleep they shared the same crystal life-support unit, Sleeping Beauty and Prince Charming in suspended animation in each other's arms.

'My older colleagues will indulge me in my choice of music,' Dr Foley begins, 'but sublimity, I would submit to you, Monsignor Frère, comes in many forms. "Good Vibrations" was one of the great achievements of popular music, so complexly constructed that even classical music critics considered it a masterpiece. It's the Rosetta stone of American mid-20th-century music and subsequently became part of the world's collective unconscious.'

Monsignor Frère nods grudgingly. 'Even in my tribal village in Nigeria when I was a boy, we sang the lyrics. We turned them into a rap: "Vibrations, yeah, baby, move smooth to the groo-oove".'

'There!' Miss O'Hara laughs excitedly. 'Eliot's point exactly.' As for Mrs Cortland, the thought of the dear Monsignor, who is on the large side, rapping and moonwalking to 'Good Vibrations' has her in a fit of merriment.

Professor Van Straaten shows his impatience. 'So have you saved the best for last?' he asks Dr Foley.

'As an historian,' Dr Foley answers, 'it is my opinion that the most transforming event in cosmological science has been the unfolding set of revelations of Earth's – and man's – position in the cosmos. Of course I shall only be able to talk of Western man and Western cosmology; in earlier times, Western cosmologists were blind to

Muslim science and astronomy and therefore to the brilliance of their Eastern elders.'

Dr Foley smiles at Miss O'Hara and proceeds with his thesis. 'In Western history, there have been three major propositions on the Earth's position and, to be frank, man hasn't dealt with any of these positions very well. Thales of Miletus, the Greek mathematician, is credited with originating the first in 585 BC, proposing an orderly universe with Earth at the centre. However, it was Aristotle, in Athens during the 4th century BC, who became the primary proponent of the geocentric model. Philosophically wedded to a vision of a universe with man at the middle, he advanced a model which established the heavens as a realm of circular perfection. Thus the sun, moon and all the planets circled the Earth in orbits that were simple, crystalline, harmonious and musical, lacking beginning or end. Then, in the 2nd century AD, when Alexandria had replaced Athens as the intellectual capital of the world, the Egyptian mathematician Claudius Ptolemy used Aristotle's template to create his own intricate model. In his *Almagest*, the Earth was still at the centre, and the stars and planets whirled in circular orbits around it; but his model also included epicycles to take into account the orbital variations shown by Mars, Saturn and Jupiter. It looked like the insides of a clock and, indeed, comprised an incredibly complicated clockwork universe.'

'Yes,' Professor Van Straaten nods. 'But with the rise of Christianity, Rome displaced Athens and Alexandria as the world's intellectual capital. Unfortunately, there, the intellectual ideas were driven by religion, not science, and, in particular, the idea that God had created the heavens and the Earth.' He casts an arrogant eye at Monsignor Frère, knowing that his remarks will ruffle his feathers. 'What the Roman Catholic church did was to combine its own ideas of an omnipotent and omniscient God with Aristotle's and Ptolemy's thinking. Theologians came up with their template: God had created man to have dominion over all. Everything in the universe was meant to serve him. Man was the centre of the Christian universe, both literally and figuratively. The church even invented an Angel of God who cranked a piece of machinery called the *primum mobile* to explain why we had 12 hours of day and 12 hours of night. It was a beautiful model of the universe: God-driven, man-driven, ego-driven. Except that it was wrong.'

Mrs Cortland casts a quick glance at Captain Craig. When he doesn't intervene, she asks Dr Foley, 'You were saying that there were three major propositions on the Earth's position? What was the second?'

'The second, Mrs Cortland, came when cosmologists, around the 15th century, suspected that the first position was wrong. It wasn't the Earth that was at the centre of the universe. Rather, the universe was sun-centred.'

'And the holy mother church,' interrupts Professor Van Straaten, 'is still struggling to come to terms with that shift, isn't it, Monsignor! Cosmology had been captured by the theologians. When scientists stated that observation and calculation, rather than divine revelation, could reveal the workings of the heavens, they posed a direct threat to church writ and threatened the proposition that God made the world and the universe – especially that the universe was eternal and unchanging, just as God had made it on the First Day. Any cosmologists who offered evidence that it might be otherwise were branded heretics. I wonder how many scientists were burnt at the stake, eh?'

Dr Foley hastens on. 'The Polish physician and clergyman Nicolaus Copernicus was the first to herald the shift. He thought the Ptolemaic system was a bit of a mess and spent much of his life trying to come up with a cleaner, simpler explanation of the motion of the planets. In 1514 he published his *Commentariolus,* putting the sun at the centre. By taking a heliocentric approach, he began what was tantamount to an astronomical mutiny. Indeed, the period from 1514 to 1690, from Copernicus through Brahe, Kepler, Galileo and Newton, has often been termed the Copernican Revolution. Galileo Galilei's appearance, of course, heralds the dawn of science. We have already heard from Professor Van Straaten at a previous dinner about that remarkable man, so I won't traverse the same ground here.'

'And the third shift?' Miss O'Hara asks.

'We have to fast-forward to 1923 for that,' Dr Foley says. 'To a certain young pioneering astronomer and scientist named Edwin Hubble.'

'Hubble, bubble, toil and trouble,' the Professor quips.

'Hubble's observations through the Mount Wilson telescope in the San Gabriel mountains above Pasadena paved the way for man to make the leap from sun-centred universe to the revelation that, actually, the cosmos was bigger than man had thought it was. Previously, a great debate had raged in which most cosmologists believed in a single-galaxy universe. One of its proponents, Harlow Shapley, thought that every star they saw was inside the Milky Way. But with Hubble came the discovery that, for instance, what was then known as the Andromeda Nebula was, in fact, a galaxy that was outside the Milky Way. In 1924, when Hubble wrote to Shapley to tell him of his findings, Shapley said, "Here is the letter that has destroyed my universe." From that moment onward, the entire universe had to be remapped and everything in it reclassified. We weren't just a one-galaxy cosmos. We were merely one luminous pinwheel among billions of similar whorls of stars. The universe was filled with countless Milky Ways, each as grand as our own galaxy.'

Dr Foley gives an enigmatic smile. 'My point is that at every revelation, ego-driven man has had to redefine himself in relation to his universe. You see, within this vastness we simply disappear. The universe is infinite, or almost infinite, going on for ever and ever. This discovery has brought us face to face with the horrifying consideration of how very small and infinitesimal we are as a species. We're not as big as we thought we were after all. Not only that, but we've had to redefine ourselves with each other. In each case, every new position has been accompanied by violence and murder, symbolised in that moment in *2001: A Space Odyssey* when a group of apes kills one of their own. A leader among the apes has used a jawbone as a weapon in the murder. Triumphant, he throws the bloodied jawbone into the air, where it transforms itself into a spaceship – just like ours.'

You, great universe,
All you suns and stars,
What would be your happiness
If you did not have those
For whom to shine?

4.

Captain Craig looks at Mrs Cortland and remembers the question she had asked before dinner: 'We're almost there, aren't we?' Uncharacteristically, she had shivered.

She has regained her calm. But there was a time – oh, how many billions of light-years ago was it now? – when the courage of all on board *Endeavour* had been tested to the limit.

That moment had come when *Endeavour* had reached its original destination: HUDF-JD2, the gigantic galaxy that lies within the Hubble Ultra Deep Field. Up to that point, all on board had thought they were simply on a voyage of discovery, a voyage which Mrs Cortland's late husband Peter had planned to lead.

'Our astronomers have detected something strange developing out there,' Peter Cortland had told Walter Craig. 'Although we suspect we know what it is, or, more importantly, what it means, I'm building a ship to travel to HUDF-JD2 to take a look firsthand. Our projections suggest that the galaxy cluster which we see forming there billions of years ago is in a region of space so dense that there will by now be a supermassive black hole – HUDF-JD2-BH1 – of some hundreds of billions of solar masses. By comparison, the 3 million solar-mass black hole in our own galactic centre is a pipsqueak.'

Cortland's eyes had twinkled with amusement. 'I'm calling the ship the *Endeavour*. There's a good reason for the name – we expect not just one gigantic black hole, but two on almost a collision course. My people have nicknamed the smaller black hole Venus II.'

Captain Craig had smiled at the parallel with Captain Cook's mission to observe the transit of Venus.

'When Venus II transits in front of HUDF-JD2-BH1 at the closest point of its orbit, the production of gravitational waves will send out the largest seismic warping event in the space-time continuum since the Big Bang itself. The back-reaction on space-time will allow us to assess the anomaly in the background geometry of the universe. We've got to be there for the transit.'

Halfway through building the ship, however, Cortland, like many others on Earth, fell ill. 'Looks like it's up to you, old girl,' he had said to his wife. 'Complete the *Endeavour* and sail it for me.'

Then he summoned Captain Craig to his bedside and gave him his instructions.

'Observing the transit of Venus II is only the first part of your mission,' he'd said. 'Once you've done that, I have a second, secret set of instructions for you to open.'

Captain Craig had smiled. He thought of the boyish games he and Peter Cortland had played during their golden-weather New Zealand childhood. 'A secret set of instructions.' Peter always had a way of talking that was talismanic, scattering his words like runes on glowing embers.

A peal of laughter interrupts Captain Craig's thoughts. He looks across the table and sees Miss O'Hara recovering from some joke that the Monsignor has told her. She turns to the waiter. 'My compliments to the chef,' she says. 'The Balinese salad was just yummy and the seven-spice Singapore duck simply delicious. Now I could dance all night! Come, partner me, Eliot.' She pulls the protesting Dr Foley to his feet, and spins him into a waltz.

Against the black velvet backdrop of the largest ballroom in the universe, Captain Craig sees a sudden shimmer and scintillation through the starboard windows. It is Aunti-2, one of the three Advanced Unified Navigational Tracking Intelligence robotic probes that accompany the *Endeavour*, on its regular parabolic flight back to the ship. Aunti-2 is soon joined by Aunti-1 and Aunti-3, and Captain Craig hears them chattering away to Hemi.

'Is everything all right, Hemi?'

'Yes, sir,' Hemi sighs. 'Aunti-3 has complained that she stubbed her toe on a moon and Aunti-1 is cross at me for sending her too far forward to check ahead.'

'There could have been Injuns out there,' Aunti-1 says. 'I could have been scalped.'

Peter Cortland's scientists have given the probes the personalities of three grumpy old ladies who are constantly scolding Hemi as if he were their nephew. In particular, Aunti-3 has been programmed to sing mid-20th-century popular songs in moments of stress.

'I'm glad you're safely back,' Captain Craig says. He knows that none of the human passengers would have survived the voyage had it not been for the aunties.

The Captain's thoughts return to that day when *Endeavour* had arrived at HUDF-JD2. Mrs Cortland still liked to keep calendar time and she'd marked the date in her diary: 13 April, 20 days before the transit.

HUDF-JD2-BH1 had lain before them, an ominous, deep dark hole with a broad layered disk, like Saturn's rings, circling the central emptiness. Vivid, molten jets of X-rays, light, electrons and protons gushed out of the poles. As Venus II began to cut a dark shape across the path of one of the jets, the fabric of space started to convulse in shivers that grew steadily larger, rocking *Endeavour* in their wake.

Aunti-3 began to sing, 'I see a bad moon rising! I see trouble on the way!' She had hovered uncertainly while her braver sisters zipped closer to take a look.

'Peter was right,' Professor Van Straaten had said to Dr Foley. 'There's no doubt, from the way that the gravitational waves from Venus II are shaking us around, that what he suspected back on Earth can be confirmed.'

'Let's wait for what the aunties have to say,' Captain Craig replied.

Very soon, Hemi had begun to relay their information. 'Captain Craig? The aunties report that the spectral analysis of the gravitational waves produced in that disturbance can be consistent only if the background curvature of the universe is positive. Furthermore, if we integrate the entire matter content of everything our sensors have encountered on the way here, it would appear that the density of clumped matter is above critical. The level of particle creation detected from the vacuum would also indicate that any dark energy has completely dissipated. Finally, measurements of the light of distant galaxies reveal that the expansion of space is slowing down and will ultimately reverse.'

'What does it mean?' Miss O'Hara had asked.

'Sometime down the track, the red shift will turn to blue shift,' Dr Foley said.

'Blue shift? That sounds very, *very* bad.'

Dr Foley had smiled at her uneasily. 'What it signals is that our universe will not last for ever. While I was working for Peter Cortland we had our suspicions. When he was dying I was the only one of his scientists he would trust to send out here to confirm them.' He turns to Captain Craig. 'He told me that he had left you a further set of instructions, Captain. Perhaps it's time you opened them –'

And Captain Craig had opened the letter:

Walter, if you are reading this letter you must by now be aware that the universe is closed and sooner or later will start contracting. Sorry to do this to you, old friend, but you're humanity's only hope. Go forward to zero.

Accompanying the instructions was a codicil which was not to be opened until zero had been reached.

5.

Dr Foley pulls a reluctant Miss O'Hara back to the dinner table.

Captain Craig turns to Professor Van Straaten. 'Professor, it's your turn to tell us what you consider to be one of the transforming moments in cosmology, is it not?'

Spuntato ecco il di d'esultanza
Onore ai piu grandi dei Regi!
The happy day is filled with joy
Honour to the mightiest of kings!

The music the Professor has chosen to accompany his offering is grandiose, martial and triumphant. He has a self-satisfied expression on his face as he waits to see if any of his fellow guests can recognise it.

'I'm sure it's Giuseppe Verdi,' says Monsignor Frère. 'The triumphal scene from *Aida,* perhaps?'

'Right composer,' Professor Van Straaten answers. 'But no, the opera is *Don Carlos,* which Verdi composed in 1886 to 1867. It explores the private and political turmoil in France and Spain in 1559 after the treaty of Cateau-Cambrésis, when Philip II took the throne of Spain. The recording itself comes from 1966 and features the divine Renata Tebaldi with Carlo Bergonzi, Nicolai Ghiaurov, Dietrich Fischer-Dieskau and Grace Bumbry, all under the baton of Sir Georg Solti. Tebaldi was considered to have *una voce d'angelo.* At the time of the recording there was an obscure rival, Maria Callas, who was apparently more highly regarded, but she fell quickly out of fashion.'

'The opera and music are not familiar to me,' the Monsignor concedes.

'I will make you a copy for further reference,' the Professor says. 'You should enjoy it. The music graphically and dramatically encompasses a scene before Valladolid Cathedral. The King, Queen, royal court and clergy assemble in front of a rejoicing populace. It is an *auto-da-fé* –'

Monsignor Frère hisses with indignation.

'– the public ceremony of the Spanish Inquisition where sentences are pronounced on the paraded heretics. At the end of the scene, the heretics are burnt at the stake. I am dedicating the music to you, Monsignor.'

The dedication is a slap in the face, but Monsignor Frère maintains his calm. By his demeanour, he shows that he will bide his time.

'Our dinner party dissertations have become more difficult as our trip has progressed,' the Professor continues, 'but like Dr Foley I have tried to keep the best till last. Indeed, Dr Foley, your thesis provides a good starting point for mine. This evening, ladies and gentlemen, I offer as my *pièce de résistance* the way in which the telescope transformed the sciences and, in particular, the cosmological sciences. For instance, at a previous dinner I talked about Galileo. Should not the maker of Galileo's telescope also be credited with Galileo's discoveries? With the telescope, Galileo was able to see more stars than had ever been seen before and to subvert church assumptions; for instance, about the moon – it was not perfectly circular, a mirror to the sun, but had a face covered with cavities and prominences. And the universe was not stationary. It was moving.'

'*Eppur si muove*,' Dr Foley recalls.

Professor Van Straaten gives a benign smile. 'But the biggest breakthrough, as far as the telescope was concerned, came when Edwin Hubble charted the recession of the galaxies. Dr Foley has already referred to this moment, on 6 October 1923, when Hubble, surveying the Andromeda nebula, found a Cepheid variable – a star so distant from our own galaxy that it proved mankind did not live in a cosy, one-universe room. But Hubble also made a second discovery, one that was so astounding it caused yet another of those huge revolutions in our thinking about the universe. He saw what was known as the red shift effect – that the wavelength of the light of all the galaxies was stretched. Not only that, but the farther away the galaxy was from our own, the more the light was stretched. This meant that the galaxies were moving away from us. But it only made physical sense if the space in between the galaxies was expanding, like a balloon being inflated. This discovery of an expanding universe, at a rate that was named the Hubble constant, led to enormous scientific curiosity about why our universe was behaving in this way. Ultimately, that led to the question of how.'

'The Big Bang,' Captain Craig murmurs. 'The very moment of creation.'

'The answer was not long in coming,' the Professor continues. 'In 1925, a few years before Hubble announced his discovery that the universe was expanding, Georges Lemaître, a Belgian mathematician, hypothesised a time when the universe might have been as small as an atomic nucleus. He studied detailed mathematical models which built on solutions to Einstein's equations first written down by a Russian, Alexander Friedmann, in 1922. Lemaître proposed that the origin was "a cosmic singularity, a day without a yesterday". He also suggested that not only did space begin with the Big Bang; time did too. Thus the unfolding of the universe was also the unfolding of time and space – the space-time continuum.'

Professor Van Straaten takes a sip of his wine. 'But Lemaître's hypothesis flew in the face of the current scientific consensus, which was that the universe was infinite, eternal and static. Very soon, scientists began to take sides –'

'Big Bang versus Steady State,' Dr Foley remembers. 'It was a huge question, very controversial. Albert Einstein, who started it all, changed his views during his career. In 1917, before Hubble's discovery of the expanding universe, his theory of relativity naturally gave a universe in flux. To cover up this embarrassment, Einstein added a finely-tuned cosmological constant to his equation, thereby hoping to stabilise his theory and continue its compatibility with the prevailing consensus of a changeless universe. But when Hubble's discovery was reported to him, Einstein realised that events had overtaken his views. He declared the cosmological constant to be his greatest mistake.'

The Professor nods. 'The irony is that scientists have subsequently argued over the cosmological constant! Mistake? Or yet another of Einstein's magnificent contributions? Dr Foley, you yourself discussed Einstein's great contributions to mankind at our first dinner.'

'E = mc²,' Dr Foley says. 'There's no doubt that Einstein was the greatest scientist of the 20th century. From his *Theory of Special Relativity* in 1905 through his *Theory of General Relativity* in 1915 he changed the way we looked at time and space – and he was still working on his *Theory of Everything* on his deathbed. Einstein was a paradigm of clarity but he could be wrongheaded in endeavouring to maintain his own views.'

Professor Van Straaten agrees. 'Although his theories paved the way for quantum mechanics, statistical mechanics and cosmology, when other scientists like Werner Heisenberg and Niels Bohr overtook his research by inserting uncertainty, unpredictability and probability into the equation – the possibility that in mathematics, at the subatomic level, behaviour could be bizarre and capricious – he kept to his own, obstinate beliefs. He was wrong. However, Einstein did support the idea of the expanding, evolving universe for most of his career. In contrast, Fred Hoyle was opposed to the idea and it was actually he who coined the derisory term "Big Bang". In 1948, Hoyle, along with Hermann Bondi and Thomas Gold, proposed a compromise which gave hope to those who still wanted an eternal universe. They theorised that the universe as a whole stayed the same, with new matter being continuously created at a slow rate, even as the individual galaxies moved away from one another and died. At the same time, proponents of the Big Bang such as Robert Dicke, George Gamow, Ralph Alpher and Robert Herman were busy calculating its consequences, such as relic radiation – the cosmic microwave background – which would still fill the universe today.'

'The Big Bang won out,' Dr Foley says. 'The evidence was too overwhelming. No scientist today doubts that there was, billions of years ago, a flash of fire. A massive explosion. All the mass and energy in the universe was created in that core moment, as was the space-time continuum.'

'This continuum,' the Professor continues, 'born out of spectacular and violent energies, inflated rapidly during the first 10 seconds. It created a cataclysm of staggering dimension, a nuclear fusion event beyond any human imagining. Its initial components – protons, neutrons and electrons – were first a primordial plasma, a fiery soup in which light was trapped. When the universe was 300,000 years old the soup cooled to the point that the first stable atoms formed, 75 per cent hydrogen and 25 per cent helium – and the light which scattered from the last free electrons was finally able to escape. It was 3000 degrees hot back then, and it has now cooled to form a microwave hiss that is the same from every part of the sky. Part of the hiss on a TV set, the static noise between channels, is the echo of the Big Bang, the microwaves that scattered from the last free electrons.'

The Professor turns to Mrs Cortland. 'The primordial plasma was very smooth at the time of the scattering. But small ripples – which were just one part in 100,000 more dense than the background – started to form clumps by gravitational attraction that eventually formed larger and larger clumps of gas. These systems of gas bonded and broke away from the expansion of the universe to form stars, star clusters and galaxies. Our own Milky Way formed in this way about 12 billion years ago. However, the heavier elements from which we are made did not form in the Big Bang; they were formed in the furnaces of short-lived massive stars which exploded as supernovae at the end of their lives, splitting their contents into the galactic disc. The shock waves from these supernovae also started the second and third generations of stars like our own which formed about 5 billion years ago – well, 5 billion years before we left Earth, I mean! The second and third generation stars contain the heavier elements from which life is made. We are star stuff.'

Dr Foley nods. 'Over all these years since, the effect of the Big Bang has kept on creating all the galaxies and stars we both know about and don't know about and – until we saw the anomaly at HUDF-JD2 – it showed no signs of stopping. Now we do know that the expansion is not only slowing down but reversing and contracting, and soon we will actually make it to the end of the universe.'

'What will we find when we get there?' Mrs Cortland asks.

Captain Craig takes up the challenge of the question. 'The journalist and mathematician Charles Seife thought it might look like an enormous wall of radiation, a wall of fire still moving and not slowed down at all by gravity. Others have assumed a wall of ice, gradually slowing like the edge of a glacier –'

'Those views assume that there's a spatial edge to the Big Bang,' Professor Van Straaten cautions. 'There isn't an edge. As I just explained, the Big Bang was not an explosion in a pre-existing space. It happened everywhere at the same time, creating space with it. The wall of fire was the surface of last scattering: it is the most distant thing we can see, because light shows us the universe in our past, not the universe as it is now. When we reach the end of the universe, again, it will be a time rather than a place. And we may arrive for the Big Crunch.'

The Professor pauses. 'It will be spectacular. Everything will come rushing together, becoming more and more dense. The blue shifts will get larger and larger in the hottest bath of radiation imaginable. The echoes of the whole history of the universe – every movement ever made, every signal ever sent – will come crashing in as space smashes together under its own weight.'

'Some scientists,' Dr Foley agrees, 'speculate that the universe will process almost an infinite amount of information right at the end. In the initial stages of the Big Crunch it will not look quite like the beginning, since the matter will be in highly processed, clumped forms and there will be millions of black holes – quite different from the primordial soup from which it all formed . . . And yet, and yet . . . we just don't know what happens at such high densities.'

Professor Van Straaten smiles at Mrs Cortland. 'If the space-time continuum comes to an end and time as we know it stops, then perhaps the end is also the beginning. At that point, going forward may also be the same as going back in time.'

'Back?' Mrs Cortland asks. 'I don't understand.'

'Neither do any of us; we try to but there's a limit to physics,' the Professor sighs. 'In travelling to the end of the universe, we might also be travelling back to its beginning. We will soon find out.'

Could that also be happening? Could the *Endeavour*, in going forward to zero, also be going back to zero?

Surprised by Professor Van Straatan's conjecture, Captain Craig recalls the recurring dream he has been having in deep sleep.

A wild landscape in the country of his birth, New Zealand. A sudden mist has descended and he is lost in it. Then he sees three elderly women from the village walking in front of him. Relieved, he runs after them. One of them turns and asks, 'Went the day well, sir?' He nods and asks directions. 'You are almost at your destination, lad. There it is.' They point to a faraway farmhouse on the other side of the valley. 'You had better make haste,' they say. 'Night is coming and, with it, a fierce storm.'

He walks to the farmhouse. As he approaches, Mrs Cortland opens the door. 'Thank goodness you were able to make it home before dark,' she says. 'Come in, come in.'

When he enters he sees that Monsignor Frère, Dr Foley, Miss O'Hara and Professor Van Straaten are having a cup of tea and chatting.

'Let Walter have a place by the fire,' Mrs Cortland scolds, and Monsignor Frère makes room for him. 'He'll catch his death. There's room for one more.'

The five people in the room smile at Captain Craig as if they have known him all their lives. He shifts on his feet nervously. Mrs Cortland brings him a cup of hot tea. 'This will bring you back to life,' she says.

'Do I know you?' he asks her. 'Do I know any of you?'

'Of course you do!' the Professor laughs. But he is uncertain, and his laughter fades away into bewilderment. 'Because if you don't, then who are we?'

The Monsignor winks at him. 'Interesting, isn't it?'

Professor Van Straaten leans forward, reflective.

'You know, it has always been my view that all of man's greatest achievements have been in the sciences. What achievements in the humanities can possibly compete with the great advances in medicine, health sciences, earth sciences, engineering, food sciences, biotechnology and information technology, to name just a few? Yes, literature has had its Prousts and Shakespeares, music its Bachs and Mozarts, and art its Rembrandts and Picassos, but they have provided only variations on the human artistic wish to represent cultural attainment from generation to generation, nothing more. My colleagues in the humanities have often disagreed with me, but I stand by my opinion that the sciences have been where the true advances are and that they have provided no equivalent achievements. Yet with all the scientific potential we had, what *happened*?'

'Indigenous people,' Captain Craig begins hesitantly, 'would tell you that the division of the sciences and the humanities into two separate disciplines was a grave mistake. How can the head function without the heart, the body without the spirit, the individual without understanding his role in the community?'

'Let's hope Peter Cortland was right,' the Professor says. 'He believed in the multiple Big Bang model, a universe that could recycle itself. He also believed in the universe as a 10-dimensional construct.'

Professor Van Straaten's voice cracks with despair. He gestures helplessly at the universe sparkling with the eyes of Heaven.

'We had all this to play with, but we failed to heed the warnings of people like Jared Diamond, Ronald Wright and Martin Rees, among many others. Nor did we heed the warnings of history – the Te Rapa Nui experience, for example, when cutting down all the trees on Easter Island led to the destruction of that fragile environment; in 2005, only 10 per cent of the planet's natural forest was left. We kept on pushing Nature over critical threshholds and, finally, past the point of no return. We couldn't even manage our possibilities, let alone our own planet.'

6.

Was Peter Cortland right? Now that *Endeavour* had almost reached the end of the universe, would going forward to zero prove the right choice to have made?

Captain Craig recalls the arguments that raged among them after he had opened Peter Cortland's secret instructions. Stationary above Venus II, he and his passengers had tried to come to grips with the implications of a contracting universe, and with Cortland's belief in a universe as a 10-dimensional construct, let alone one that could recycle itself. Their courage – and their sanity – had indeed been tested to the limit.

Over the following three days they had continued arguing about what to do. One point became increasingly clear: returning to Earth was not an option. What was there to return for? Earth's sun would have expanded to a red giant long ago, engulfing the Earth tens of billions of years before. Still growing, the sun – if it had survived the collision between the Milky Way and Andromeda – would itself have died in a cataclysmic explosion that left nothing but a cold, dense white dwarf.

But what did the final death of the Earth matter to humanity? On the point of *Endeavour*'s departure, most of the world's population was dead. Everyone. Gone. All condemned to death by ecological collapse or by the pandemics that ravaged the planet and lethally crossed the species barrier. Why should it have been such a surprise when the collapse of the biosphere occurred? Mankind, with all the innocence of a child playing with ingenious toys, had caused it itself. How was anybody to know that the future would be so frail?

And so, as had been predicted in the *Mahabharata*, the time of Kali had truly settled to Earth. Countries tried to stockpile resources on the one hand and to put up barriers to stop the viruses from spreading on the other. Behind national barricades, they tried to maintain law and order. When that failed, the survivors sought refuge from a poisoned planet by retreating to countries like New Zealand at the Earth's perimeters.

But all succumbed, all died. All life, not just humankind. Captain Craig's own wife and two infant children, once vital and full of life, gone, wasting away, their beauty blighted by viral infection. Even Peter Cortland, while completing his ship within the shadow of Mount Hikurangi, the first place in the world to see the morning sun – gone. But not before he had given his childhood friend Captain Craig his enigmatic instructions – *Walter, go forward to zero* – and the codicil to be opened when zero had been reached.

There was no doubt that Peter Cortland had planned for an ongoing voyage. Why else had he built *Endeavour* for time-dilation travel? Leveraging off his military contracts with the US government, he had secured antimatter supplies for his own nanotechnological corporation. Working with the most advanced technology available, his scientists had engineered the Cortland photon epsilon drive so that *Endeavour* could travel the staggering dimensions of the universe. It was his scientists who, working from the solution to Einstein's equations discovered by New Zealander Roy Kerr, perfected the techniques by which *Endeavour* was able to use black holes to slingshot its way across the space-time continuum. Only with such a drive could *Endeavour* make it to the end of the universe.

And why else, during the same time as the epsilon drive was being tested, had Cortland had another team of scientists working on a life-support system which could sustain human life over billions of years in intense radiation fields – and maybe, if Cortland's beliefs proved wrong, even in a Big Crunch? When *Endeavour* was not travelling at relativistic speeds, all the passengers had to be stored for extended periods in suspended animation. Whether sleeping or awake, they also required oxygen supplies and artificial gravity, produced by rotation of the habitable spaceframe. Despite *Endeavour*'s immense size, the vessel had been designed to take just seven people: Cortland himself had planned to be one of them. They were to be the only humans on board. The ship itself and all service crew were robotic, controlled and directed through Hemi's Artificial Intelligence matrix. The three aunties had been added to the design specifications just before Cortland died.

In the end, it came to a simple vote. Find a habitable planet and try to remake an Earth for all those sentimental reasons that afflict the human family? Or keep going to the end of the universe and hope that when you got there a solution would present itself?

Peter Cortland's vote was already clear. And so was that of Mrs Cortland. 'Let's go forward,' she had said. 'Even if Peter is wrong, we would still have a chance of making it through the Big Crunch, wouldn't we? If so, we could still find that habitable planet –'

Captain Craig looked at the other passengers. They had nodded in agreement. Whichever way you looked at it, any survival option was slim.

'In that case, Hemi,' Captain Craig instructed the avatar, 'full speed ahead.'

'As you wish,' Hemi had replied.

And he headed *Endeavour* towards the monster black hole rotating at the centre of HUDF-JD2. Shepherded by the clucking, fretting aunties – 'Do we have to go there?' Aunti-2 complained. 'Cold, dark, windy' – they entered the swirling vortex.

The interaction of rotation and huge gravitational and magnetic fields did the rest. The ship was accelerated to huge speeds and deflected out rather than being sucked in. Before Captain Craig and the passengers knew it, *Endeavour* had been kicked forward and over the perimeter of the known universe.

Captain Craig remembers the silence that had suddenly descended. HUDF-JD2 had been the last tangible reminder of home. From now on they were travelling into the unknown, just like Cook had done those many centuries ago.

*

'Come now, Monsignor,' Dr Foley laughs. 'Don't let Professor Van Straaten put you off your stroke. Your turn now.'

With a start, Captain Craig comes back to the present.

'My learned friend,' Monsignor Frère begins, nodding at the Professor, 'has been at pains to underline the difference between theology and science, to theology's detriment. But the church acknowledged its mistakes and pardoned Galileo –'

'What a pity it took so long to do it,' the Professor says. 'His *Dialogue on the Two Chief Systems of the World*, from 1632, was placed on the *Index Librorum Prohibitorum* and stayed there for 200 years.'

'Please be generous, Professor,' the Monsignor responds. 'Doesn't science forgive itself for its mistakes? Allow yourself the generosity of also forgiving theologians. We may have been on different sides once, but no longer.'

'You still believe that God made the universe!' Professor Van Straaten exclaims. 'We are as irrevocably opposed as we ever were.'

'Even men of religion,' Monsignor Frère continues, 'interrogated the church's ideas on cosmology. Cardinal Nicholas of Cusa in 15th-century Germany was one such man. Not only did he dispute the concept of an Earth-centred universe, he also challenged church dogma by hypothesising the existence of other Earths with other moons – and all inhabited by their own intelligent kind. This was a doctrinally dangerous position for any cleric, as it questioned the role of God, the Vatican as the One True Church and the entire purpose of the universe as having been made for the specific benefit of man.'

'You are painting your church in a benign light,' the Professor rejoins. 'The entire cosmological debate since the Middle Ages has been one of conflict between church and science. For instance, although Copernicus supported Cusa's doctrine he delayed publication of his own scientific findings – *On the Revolution of the Heavenly Spheres* – until 1543 and the very moment when he was beyond the reach of the church: on his deathbed.'

Monsignor Frère sticks to his guns. 'And to speak of the Galileo matter,' he continues, 'in 1992 Pope John Paul II spoke against the myth that had arisen from the church's treatment of Galileo and "the church's supposed rejection of scientific progress". It was not rejection but "tragic mutual incomprehension". Galileo himself would be pleased that, now, the Vatican Observatory embraces the scientific method. Just as scientists have learnt, we have learnt. The Bible has taught us how to go to Heaven, not how the heavens go.'

Turning to his fellow passengers, the Monsignor expands his appeal. 'Professor Van Straaten also conveniently forgets that in 1951 Pope Pius XII endorsed the Big Bang model. At the Pontifical Academy of Sciences that year he gave an address, *The Proofs for the Existence of God in the Light of Modern Natural Science*. To wit: "Thus everything seems to indicate that the material universe had a mighty beginning in time, endowed as it was with vast reserves of energy, in virtue of which, at first rapidly and then ever more slowly, it evolved into its present state . . . In fact, it would seem that present-day science, with one sweeping step back across millions of centuries, has succeeded in bearing witness to that primordial *Fiat lux* uttered at the moment when, along with matter, there burst forth from nothing a sea of light and radiation, while the particles of chemical elements split and formed into millions of galaxies . . . Therefore, there is a Creator. Therefore, God exists! Although it is neither explicit nor complete, this is the reply we were awaiting from science, and which the present human generation is awaiting from it."'

'We are not on the same side,' Professor Van Straaten mutters. 'The Pope's address was a thinly disguised attempt to align the Big Bang with the Book of Genesis. Your people manipulated scientific evidence for your own ends to persuade your flocks that there was no dichotomy between science and religion. Even your own scientist, Monsignor Georges Lemaître –'

'The inventor of the Big Bang model,' Monsignor Frère emphasises. 'A Jesuit priest and previous head of the Pontifical Academy of Sciences – a point you conveniently glossed over when you were talking about him –'

'– firmly believed that science and theology should stand apart from each other.'

The Monsignor winces. 'Please do not quote one of my colleagues at me. I would remind you that Monsignor Lemaître made clear his position: "To search thoroughly for the truth involves a searching of souls as well as of spectra. Scientists would be wise to do both."'

'The church lost,' the Professor persists. 'Scientists know how the cosmos began, and God had nothing to do with it. Despite the persistence of ideas about an eternal, unchanging universe, the evidence that scientists have collected points to a cosmos that was born and which, at some point, will end – and God cannot change that. Nearly two millennia of philosophy and theology were replaced when science triumphed over religious superstition. Where, Monsignor, is the evidence that God Himself actually exists? Wouldn't you have thought, with all of our scientific observations and calculations, that man would have seen Him by now? Where is He hiding?'

'Gentlemen, please,' Mrs Cortland intervenes gently. 'Monsignor, what are you offering tonight as the most transforming event in the history of science and, in particular, the cosmological sciences?'

The Monsignor calms down. 'So far,' he begins, 'Dr Foley and Professor Van Straaten have told us how the universe was discovered and what lessons were learnt about the seen universe. I would like to posit that, from the 1990s onward, the most transforming event in the history of the cosmological sciences was the discovery of the *unseen* universe.'

Professor Van Straaten gives a small laugh of surprise. 'The time has come, the walrus said, to speak of many things,' he sings, 'of baryons and gluons, and quarks and other things.' He nods grudgingly. 'Yes, yes, you surprise me, Monsignor. Well done.'

'If I may continue?' the Monsignor smiles. 'By the beginning of the millennium, apart from better telescopes, science had developed super-precise tools – cosmic microwaving, fibre optics, interferometry, spectrometry, cosmic background imaging, the super-proton synchrotron, relativistic heavy-ion colliders, particle accelerators, charge-coupled devices, super-symmetry, microlensing and so on. These developments enabled cosmology to enter an era of stunningly precise measurements. The major problem was that when the cosmologists tallied up all the matter that they could *see* – everything in all the galaxies, stars and planets – there just didn't seem to be nearly enough stuff to make a universe. Not enough baryonic and leptonic matter – protons, leptons, neutrons, electrons, mesons, pions, muons, tau particles, antimatter, other antimatter twins, quarks and gluons –'

'Is the Monsignor being serious?' Miss O'Hara laughs. 'They sound like creatures in a fantastic zoo.'

'Indeed I am, young lady. But more surprises were to come, because when the scientists completed their tally of the stuff that was visible they had to face up to the fact that at least three-quarters of the cosmos was composed of material that humans could *not* see and had not directly measured, yet it must be there!'

'The Case of the Missing Matter,' Professor Van Straaten grunts. 'Yes, yes.'

The Monsignor is clearly enjoying himself. 'So how do you find this missing stuff if the world's most sophisticated telescopes can't even see it? Well, this was where the new, super-precise measuring equipment came in. Instead of looking for the missing matter directly, scientists began to look for it indirectly – not at what it looked like, but at what it *did*. Their best shot was to use tools that weighed the universe's mass and measured gravitational attraction – the mutual pull between two objects that have mass. Finally, they found what they were looking for.'

'They called it non-baryonic matter,' Dr Foley explains. 'Sometimes it was referred to as exotic matter, but most often it was termed dark matter. The suspicion was that this dark matter could be found in one of those zoo animals you referred to, darling – neutrinos, one of the most elusive particles known to science. The structure of the galaxy clusters, the motions and distributions of the galaxies, and the fine details of the cosmic background radiation all implied that the universe was filled with it.'

'But,' the Monsignor continues, looking out *Endeavour*'s windows, 'it appeared that there was something *else* out here –'

'Dark energy,' Dr Foley says. 'A strange anti-gravity force that functioned ironically and suspiciously, very like the cosmological constant Einstein had self-discredited years before. It counteracted the effects of gravity by keeping up an outward pressure which for some billions of years had been strong enough to accelerate the expansion of the universe. But nobody could agree on what the dark energy was: whether it would stay constant and make the expansion of the universe accelerate for ever, or whether it would dissipate, allowing the universe maybe to contract. There were those who proposed a phantom energy which, while violating some basic physical principles, would make it possible for the universe to end suddenly in an unannounced Big Rip rather than a Big Crunch.'

'Yes,' Monsignor Frère nods. 'Whatever it is, this strange anti-gravity force has been the greatest mystery in our understanding of the universe for most of the 21st century.' He purses his lips. 'And if we were calculating its density against the total stuff – mass and energy – in the universe, it would actually comprise by far the greater component.'

'Greater?' Mrs Cortland repeats, surprised.

'Which brings me to my point,' the Monsignor muses. He looks steadily at Professor Van Straaten. 'We now know since we passed HUDF-JD2 that the dark energy, if it was there, must have dissipated, making it a decaying quintessence field. But was the dark energy ever real, or were the cosmologists who suggested that we had misinterpreted distance measurements in a universe with large-scale lumpiness actually correct? Interpreting cosmological measurements is a complicated forensic science. It depends not only on what is actually measured, but also on the assumptions about the physical laws, and the geometry of the universe, that are used for interpreting those measurements. One can never recreate the Big Bang in the lab and repeat the cosmic experiment. You know that, Professor, and I know that. And there are so many models – inhomogeneous universes as well as "quintessence" models. We didn't know everything when we left; do we finally know the model of the universe now?'

Monsignor Frère again expands his address to include the audience. 'I would have to say that, regardless of my colleague's protestations, the presence of the unseen affirms the continuing presence of God in the universe. I knew it when I was a boy staring at the night sky from my small African village in Nigeria, and I know it again tonight as we dine aboard the *Endeavour*. The universe is not a vacuum; it is, rather, a cosmological gene pool.' He begins to hum: 'All knowing, all loving, invisible God –'

'Oh no you don't,' Professor Van Straaten interrupts. He knows exactly where all this is heading.

'Despite all the scientific advances in cosmology,' Monsignor Frère continues, 'humanity still believes in a higher purpose for itself and the universe it lives in. Professor, earlier you asked a question, "Where is God hiding?" He hasn't been hiding at all. Nor, so Albert Einstein believed, was He hiding anything from us; He was just asking us to search harder. Einstein thought of God as a gardener. Einstein was simply trying to catch Him at his work. So it's all a matter of where to look and how to look – not just with your head but also with your heart, not just with your instruments but with your intuition, not just with certainty but also with faith. This may be an enigma for you, Professor Van Straaten, but for people of faith, the answer is simple. If you wish to find God, eternal, unchanging, you need only seek him in the invisible and the ways in which He continues to maintain the space-time continuum.'

'You're clutching at straws,' Professor Van Straaten mutters.

The Monsignor gives a sigh of regret. 'What a pity, Professor, that the relationship between scientists and the rest of us – humankind as a whole – has always been so . . . so adversarial.'

'You're as much to blame for that as we are,' the Professor answers. 'The church was the main agent in maintaining the deep conflict about the nature of knowledge and the rules which governed its use.'

'Yes, my friend,' Monsignor Frère says gently, 'perhaps. But I am honest enough to regret that we could not find a competent way of working together – you, me, leaders of governments, corporate leaders – before it was too late. We should have developed an international meritocracy to override the petty proliferation of national communities. All wanted their own nuclear bombs and their own research capabilities designed specifically to enhance their own financial, political and economic growth. The world grew out of kilter so quickly. It wasn't that we failed to heed the warnings, Professor. We just couldn't find the ways to deal with them as a planet. Man's own volatile nature sealed his destiny. Why did everything in the human world tend to disorder?'

Fly toward the Lord
Speed, oh, you poor souls
Come to your salvation
Kneel before the throne of God

At that moment, the ship rocks.

Captain Craig hears Hemi give a surprised gasp. 'Yes, Hemi?'

'Captain, my fault entirely. Something extraordinary has occurred. It completely put me off my stroke. But I've restored the ship's trim. Aunti-3 has picked up some kind of transmission that can't be accounted for. Please convey my apologies to your dinner guests. But perhaps you and I could talk privately?'

'I'll come to the bridge,' Captain Craig answers.

'Impeccable timing,' Mrs Cortland whispers to him behind her hand as he stands and excuses himself. 'This will allow the table to be cleared for dessert, and for Miss O'Hara and me to repair to the powder room. Let's hope, by the time you get back, these silly men are over their irritation with one another.'

The *Endeavour* has three decks. Captain Craig makes his way to the bridge on the uppermost tower, where an array of monitors shows views fore and aft, and continuous computerised read-outs from Hemi's mainframe.

He seats himself at the central console. 'So what is it, Hemi?'

'I have received a message for you.'

'A message? That's impossible. Everyone on Earth is dead.'

'The message does not come from behind us. It comes from in front of us.'

'Then put the message on the screen,' Captain Craig orders.

The screen shows a mathematical equation:

$$\sup\{|r(t) - r_\oplus|\} = |r(t) - r_\oplus|\,!$$

'Give me a translation, please, Hemi.'

STOP. COME NO FURTHER.

'What does it mean?'

'There are two possibilities,' the avatar says. 'It may be advice that we have reached our destination. Or, alternatively, it may be a warning to go back.'

The Captain is thoughtful. The minutes tick by. Then: 'Have we reached zero?' he asks.

'Not quite yet, Captain. We should do so within the hour.'

Captain Craig makes a decision. 'In that case, remain on course.'

7.

'Moon river,' Aunti-3 sings, 'wider than a mile –'

The ship is like a silent celestial angel in solitary flight through a sea of stars. It cleaves through the blackness, serene and powerful, its light-wings at full extension, accelerating through the space-time continuum.

Once it had left HUDF-JD2, *Endeavour* had dropped off the map. An interesting conundrum given that, in fact, all the maps of the known universe were themselves wrong or, rather, obsolete, even before *Endeavour* left Earth. Dr Foley had explained it to Miss O'Hara in this way:

'Our maps of the universe really charted points of light which had taken years to reach our Earth. Thus, Alpha Centauri was not as it was seen "now" but as it had been four years before. Similarly, as seen "today" the most distant galaxies appeared to us as they had been billions of years ago. Imagine Captain Cook setting off on his voyage to the South Pacific – the analogy would be giving him a map of Gondwanaland 350 million years ago to enable him to navigate his way.'

'So all along we've been navigating without a reliable map?'

'Yes,' Dr Foley had answered. 'Thank God for the aunties!'

Guided by the aunties, the *Endeavour* had maintained full speed through the uncharted waters of the cosmos. Half the volume of the universe was in deep voids, so her voyage was not too dissimilar to her namesake's, out of sight of land, week after week. Hemi would have preferred to point *Endeavour* at empty sky and avoid galaxies like the plague but, knowing that the humans on board craved light and 'landmarks' – and, of course, that the ship needs Kerr black holes to maintain its energy source – he and the three aunties inevitably plotted a course which was populous with galaxies and their separate discs of stars.

The trouble was that navigating this crowded route brought the greater dangers. 'Better not go there,' the aunties would report to Hemi if violent quasars mushroomed ahead. 'The bogeyman may be waiting.' When intense bursts of star formation generated galaxy-wide superwinds, Aunti-3 was prone to sing 'Somewhere over the Rainbow' and complain that if bluebirds could fly over rainbows, why couldn't they fly over superwinds. Aunti-1 and Aunti-2 would just roll their eyes and advise Hemi, 'Batten down the hatches and reef those sails, boy, we're in for a big blow!'

Sometimes, when danger was unavoidable, as when two galaxies collided or massive stars collapsed in a shower of gamma rays, the aunties would simply order Hemi, 'Put on your skates, nephew, and let's get the hell out of here.' Their jocularity masked the sophistication of their measurements. They carried an array of optical, infrared, gamma-ray, spectrographic and other sensors so that Hemi could compute the dangers, raise or lower radiation shields, change course and continue to plough ahead through the reefs and island universes that dotted the celestial sea.

'What sights have the aunties seen,' Captain Craig has often wondered, 'while we were sleeping? What dangers have they ferried us through in this voyage to the end of the universe?'

He imagines that most of the time they would simply have got on with the collection of scientific data along the way. In this manner they were no different from Captain Cook and his crew on the original *Endeavour*'s voyage to Tahiti. They took soundings, photographed their discoveries, and mapped and catalogued their galactic findings. They scorned the lack of imagination shown by human starmakers who had attached cartographic numbers like NGC55, NGC253, M82, M81, M83, Centaurus A, M101 Pinwheel, M51 Whirlpool and M104 Sombrero. They preferred more exotic nomenclature like Hikurangi Gloriosa or Marama Sublima or Ariki Imperatrix; and on one occasion Aunti-3 suggested calling a particular chain of galaxies Vagina Splendida.

'You can't call it that,' Hemi said.

'Why not?' the aunties asked. 'Doesn't the galaxy look like that to you, nephew?' The aunties were, of course, teasing him. It was always a moment of triumph to them when they dropped their data into a small winking beacon, '*We're still alive*', and set it adrift in their wake. They knew the gesture was futile; the beacon was like a small bottle with a message in it that nobody would ever find.

During the early part of the voyage the universe was astoundingly beautiful. Proudly, the aunties had filtered through the dreams of the sleepers their marvellous discoveries. They showed them planets and satellite worlds of extraordinary magnificence, with interstellar rainbow clouds drifting between. There were worlds with fractured canyons of ice, thousands of miles deep, the shards refracting the light as brilliant as the star of Bethlehem. Some of the worlds had nitrogen atmospheric conditions, emitting enshrouding mists that glowed with reflected light from their sun systems. Others were sulphur-dominated, sending huge kaleidoscopic jets of violet and emerald gas into the stratosphere, like wings of angels a million miles wide. Many had ring systems of small moons captured by the planet's gravity, or countless mini-moonlets, each a dazzling corona. There were stars – so many, many billions of stars – of all brilliances and kinds, binary, multiple, white dwarfs, neutron stars and pulsars. Spiral stellar arms wheeled like sparkling, jewelled gateways to Heaven.

All the while, Hemi and the three aunties kept the prime imperative. 'Oh, precious ones,' they whispered in soft, loving voices to the humans suspended at the brink of consciousness. 'It is our privilege and our honour to carry you.' The sleepers were like pharaonic passengers, embalmed aboard a glowing sunship. 'Royal ones, rest, be at peace, and know your journey is one that the stuff of dreams is made of.'

The aunties were careful, however, not to subject the sleepers to the moments of danger they had faced. The slightest emotional disturbance could upset the careful balance of the chemicals that flowed through their bodies as they slept. And so they were not told of the solar flares that emitted infrared radiation which threatened the ship's insulated hull cladding and protective aerogel layers. Hemi and the three aunties kept their own counsel when protostars collapsed, creating violent galactic cyclones. While the sleepers slept on, the avatar desperately maintained the protective shields that kept all on-board life-support systems operational. When huge gaseous solar storms, heated to nearly 2 million kelvins, suddenly erupted into huge photospheric tsunami, Hemi and the aunties whispered soothing sounds of reassurance to the sleepers; engines already at overload, they desperately tried to override any malfunction. Sometimes there were asteroid belts, minefields a million miles wide, to avoid: any false move or collision could strand the ship for ever. At other times Hemi had to blast a way through with the light armaments carried for the purpose: the dorsal particle beam gun or the ventral weapons system.

Then there were the supernovae giving birth to Kerr black holes.

As the giant stars' cores collapsed, the black holes would condense and begin to spin, sending out bursts of gamma rays in some of the most violent and energetic events in the universe. The explosions from a million suns in supernovae, like cosmological nuclear reactors, could incinerate the ship in a millisecond. So too the gravitational pull of black holes, planets, moons and stars constantly threatened to take her into a death spin. But *Endeavour* had prevailed.

When the dangers had been negotiated, Hemi anchored the ship in the lee of a benign star system. There, assisted by the three aunties and the robotic service crew, he patched the sails, repaired the shielding, and serviced and tested the photon epsilon drive. Once he was satisfied that all systems were go, he would hoist anchor again. The sails would unfurl, the *Endeavour* would quickly come to cruise speed and, far ahead, the three aunties would broadcast to the universe:

'Make way. Make way. Let us pass.'

Once the ship moved past the 700 billion light-years mark, the character of the universe changed. It had not yet begun to contract significantly, and the deep voids multiplied: in them were nothing but tenuous, low levels of hydrogen plasma. The voids were like huge crevasses – and *Endeavour* traversed them in absolute darkness.

Very soon, even the galaxies and stars seemed to wink out. Although one could never really ascribe human emotions to Hemi and the aunties, their behaviour betrayed an increasing nervousness. Ironically, it was *they* who now sought the light. They hated the total darkness of the deep voids but knew with utter and fatalistic certainty that they had to be negotiated. 'Will we ever see the light again?' they asked.

And all the voids seemed to be channelling the ship towards some wide primordial Night – Te Kore – of the deepest blackness. When they finally reached it, a chasm almost 72 billion light-years wide, Captain Craig remembered the Maori village of his birth. There old women of the tribe had kept their oral traditions by chanting to him as he slept in their arms:

'At the very beginning of Time,' they sang, 'was Te Kore, the Void. Then out of Te Kore came Te Po, the Night, the Long Night, the Dark Night, the Night All Powerful, the Night Without End, Te Po Nui, Te Po Roa, Te Po Tangotango, Te Po Kerekere, Te Po Tiwha, Te Po Kitea . . . '

Even in Te Kore, Hemi and the aunties tried to keep everyone's spirits up. Aunti-3 took to singing a rousing 'Yellow Submarine', and Miss O'Hara would join her in the chorus. Sometimes Hemi would illuminate the ship. He could not do this often, however, as it meant draining *Endeavour* of its epsilon-drive antimatter resources. In fact, he completely miscalculated the distance to the exit, and the ship became completely becalmed. The only way to get out of the mess was for the aunties to roll up their sleeves and use their tractor beams to pull *Endeavour* through the never-ending blackness.

As if that was not enough, by the 760 billion light-years mark the contracting universe began to create havoc within Te Kore. Gravitational wave shocks from colliding black holes stretched and compressed space.

'Better squee-eeze through, boy,' the aunties would say, 'and be quick about it.'

Hundreds of billions of light-years further, as the contractions of the universe quickened, the voids shrank more and more. The galaxies that clustered around them began to crash into each other, their light exhibiting all the ominous signs of blue shift. Hemi and the aunties gave them a wide berth, choosing the safe route through the shrinking voids. 'If it's blue it's dangerous stew,' they'd chorus. Sometimes, without warning, huge jagged warps would rent the fabric of the universe, tearing it to shreds.

The *Endeavour* was a valiant star waka, glorious to behold. It was like a small, glowing seed pod, a dandelion, in that huge dark immensity. And Hemi and his three aunties were prepared to go down fighting to ensure that the precious ones within would be delivered unto their rightful destiny.

Go forward to zero.

8.

On his return to the dinner party Captain Craig is relieved to see that equanimity has been restored to the occasion. Indeed, Miss O'Hara is doing astrological readings.

'You're just in time for me to read your stars,' she tells him. 'You're an Aquarian, aren't you? In that case – aha, there appears to be a struggle for pole position taking place, and the sextile between the sun and powerful Pluto may be felt much earlier. Uranus, the planet of experimentation, is making waves, and risk is in the air. A calculated risk is a good thing, of course. But make sure you have a plan first.'

She turns to Dr Foley. 'Eliot? Would you like to be next? You're a Capricorn, yes? Your moon enters the indecisive sign of Pisces on Friday, and there may be a period when it's genuinely impossible for you to make up your mind about a certain individual whom you're emotionally attached to.' She says this with wide-eyed innocence. 'However, do not procrastinate, Eliot. Saturn will soon start its epic voyage through the sociable sign of Leo. This is the time for you to show your hand so that the person whom you admire, although remarkably different from you, may join you and you can both advance together. Professor Van Straaten? Oh, you are going on a long trip.'

The others laugh.

'As a Cancerian,' she continues, 'planets pass through your travel zones. However, you are in for a bumpy ride. Blame it on the persistent presence of Sagittarius.'

Sagittarius happens to be Monsignor Frère's sign. He takes the dig in good grace but he can't help taking a poke back. 'My dear,' he says, 'give up astrology. We're already well across the universe and your star signs are all irrelevant history.'

'Yes,' Professor Van Straaten retorts. 'Just like God.'

As the guests are being seated for dessert, a beautiful, elegaic piece of music begins to play. It was Peter Cortland's favourite and Mrs Cortland has chosen it in his memory. But it also brings home to her the immense nature of the voyage. The stars are everywhere and the darkness between is filled with a terrible and lonely beauty. She begins to shiver. She remembers how just before dinner she had summoned up enough courage to finally ask Captain Craig: 'How long have we been travelling, Captain?'

He had given her a long look.

'Come on, Captain. I'm a big girl now. And I do own the *Endeavour*. How long has it taken us.'

Captain Craig had nodded. 'I'll ask Hemi to compute the time for you.'

The avatar had replied almost on the instant. 'Taking into

account leap years, 1401 billion, 537 million, 656 thousand and 748 years, Mrs Cortland. But with time dilation and sightseeing stops, you have aged just one year, 5 months, 21 days and 3.8 hours – though, with the extended sleep times in between, elapsed ship time is greater. Still, even that is only 27 million, 543 thousand, 81 years, 6 months, 15 days and 23.7 hours.'

Only? Mrs Cortland had almost fainted. *Oh, let's turn back. Oh please. Turn. Back.* But there was no *back.* There was only *forward.*

'What beautiful music,' Monsignor Frère says now. 'So soothing after the Verdi.'

Mrs Cortland regains her demeanour. 'It's a baroque aria, "The Plaint", from Henry Purcell's *The Fairy Queen.* The opera is based on Shakespeare's *A Midsummer Night's Dream*, and was first performed in May 1692 at the Theatre Royal, London.'

'Who is the singer?' Miss O'Hara asks.

Mrs Cortland smiles. 'Her name is Margaret Ritchie, and she was an especially admired interpreter of Purcell. She made this recording in 1954. Her interpretation is moving, don't you think?'

'A sovereign voice,' Monsignor Frère concedes. 'A sovereign aria. So, what is your topic, Mrs Cortland?'

She suppresses a peal of laughter. 'I'm sure it will not surprise you, but it is this: one would think, from looking at the history of the cosmological sciences, that it was a male history. And indeed, as in all the sciences, cosmology was the prerogative of men. However, against all odds women have also made their impact.'

'Oh no,' Professor Van Straaten groans.

'Oh yes, Professor,' Mrs Cortland answers. 'One of the first-known woman mathematicians and astronomers was Hypatia, circa AD 370 to 415. Her father, Theon, was the last head of the museum at Alexandria, and Hypatia herself became one of the last guardians of the old Ptolemaic knowledge. She wrote a commentary on Ptolemy's work and invented astronomical navigation devices. Who knows what else she might have accomplished had she not been murdered by Christian monks during Alexandria's waning years? And all of you might extol Galileo, the revolutionary polymath-mathematician, physicist and astronomer, but never forget, gentlemen, that it was his daughter who was the most important person in his life. She was Sister Maria Celeste Galilei, and her relationship with her father was so close that his letters to her are sometimes marked in the margins with Galileo's notes, calculations and diagrams.'

Mrs Cortland looks at the other dinner guests. 'Often, where you found a man you also found a woman – like Caroline Herschel, sister of Friedrich Wilhelm Herschel of Hanover. Her brother was the most famous astronomer of the 18th century, and Caroline helped in his discovery of Uranus. She also discovered eight comets during her own brilliant career, so let nobody think that she was her brother's assistant.'

Dr Foley nods his head. 'She snatched every leisure moment for resuming some work in progress. Often she was so intensely involved she didn't even take time to change her dress. Sometimes she put food into her brother's mouth as he was working.'

'So you've heard of Caroline Herschel?' Mrs Cortland asks. 'Do you also know the story of Edward Pickering and his women?'

'Oh, Pickering's harem,' Professor Van Straaten laughs.

'In 1877, Edward Pickering became director of the Harvard College Observatory. By that time in the history of cosmology, photographing the stars had become as important as observing them, and Pickering was head of a large project of celestial photography and analysis. The thousands of stars had to be tabulated, their locations established and brightness classified. During Pickering's period, over half a million photographic plates were scrutinised by the observatory; it became the largest photographic library of the universe in existence. Who did the work? Universities in those days largely excluded women, so Pickering's staff were all young, bright, intelligent male computers. However, they were also inaccurate, careless, and often lacking in the concentration and meticulous attention to detail that was required. Pickering became so cross with them that one day, in a fit of exasperation, he sacked them all. Declaring that his Scottish maid, Williamina Fleming, could do better, he hired a team of women computers – and put Miss Fleming in charge.'

'Really?' Miss O'Hara gasps.

'Mind you,' Mrs Cortland cautions, 'Pickering paid them only 30 cents per hour, whereas he had been paying the men 50 cents. Nevertheless, this was probably the first example of the suffragette impulse in the sciences. The astonishing aspect of the story, however, is that after a while the women subverted the original brief. Not content just to harvest data, they began to analyse it and come to their own conclusions. One of the women, Annie Jump Cannon, took matters further. In 1911, realising she was cataloguing around 5000 stars per month, she decided to establish a better, more expansive and inclusive framework for classifying stars by their location, brightness and colour: O, B, A, F, G, K, M. This system was universally adopted, and in 1925 Annie Jump Cannon received an honorary doctorate from Oxford University, the first woman to be so honoured by that institution – and the first of many other honours for her too.'

'Henrietta Leavitt was another of Pickering's team of women,' the Monsignor intervenes. 'Interestingly, both she and Miss Cannon were profoundly deaf.'

'She was probably more famous than Miss Cannon,' Mrs Cortland concedes. 'She graduated from Harvard University's Radcliffe College in 1892. She discovered more than 2400 variable stars during her career – a phenomenal accomplishment. She was particularly fascinated by Cepheids and decided to focus all her concentration on a Cepheid-hunt within the Small Magellanic Cloud. She came up with 25 Cepheid variables, plotted a graph of the apparent brightness against the period of variation for the 25 stars – and, in 1912, revealed that the mathematical formulae she had used to measure the distances between the Cepheids could also be applied to measure the distances between any objects in the universe. It was a decisive discovery but Miss Leavitt did not receive recognition for it.'

Professor Van Straaten gives a little shameful cough. 'Professor Gösta Mittag-Leffler of the Swedish Academy of Sciences began the paperwork process for a Nobel Prize nomination. But he was too late. Leavitt had died of cancer three years earlier, when she was 53.'

'She wasn't the only one to be bypassed,' Mrs Cortland says. 'In 1967, Jocelyn Bell, a radio astronomy student at Cambridge University, detected regular pulses in radio transmissions that turned on and off rapidly, like a broadcast being made by some alien intelligence. She denoted it as an LGM, a Little Green Man. In fact, what she'd observed was a pulsar, a new type of pulsating star. And, ironically, it was her adviser, Antony Hewish, who in 1974 was awarded the Nobel Prize for the discovery.'

'In New Zealand,' Captain Craig interrupts, 'we had young Beatrice Hill Tinsley, who did all her scientific research in the United States. In 1975 she postulated that galaxies change as they evolve, and she constructed computer models to explain the process. When she died of melanoma shortly after her 40th birthday she was a professor at Yale and had made scientific advances worthy of a lifetime's effort.'

Dr Foley nods his head. 'For such a small country, yours has certainly given the world some great scientists – Ernest Rutherford, William Pickering, Maurice Wilkins, Ian Axford, Alan MacDiarmid and Roy Kerr among them. Karl Popper also? And then, of course, Peter Cortland –'

Mrs Cortland turns to Monsignor Frère. 'Monsignor, you have mentioned dark matter. Did you know that some of the most crucial discoveries about dark matter were made from the rotation rates of galaxies by Vera Rubin? Nor must we forget women like Wendy Freeman who, in 1999, led the scientific team which completed the Hubble Key Project.'

All of a sudden Mrs Cortland thinks of her poor, long-dead, diseased Earth and all of humankind gone, gone, gone. 'I sometimes think how much better our future might have been if women had been allowed to take control of our destiny,' she says. 'I think we would have made a much better job of it. We certainly wouldn't have done any worse, don't you agree?'

And now it is Miss O'Hara's turn.

'This is so unfair,' she wails. 'You're all older than me and wiser, and I don't know anything about cosmology! Still, I'll try this my way. Roll cameras,' she instructs Hemi. 'Standby and action. Aunti-3? Take it away –'

'When you wish upon a star,' Aunti-3 sings in a lovely lyric soprano voice.

'Maybe she'll crack on a high note,' Aunti-1 and Aunti-2 grumble hopefully. But Aunti-3 sings on, oblivious, and the interior of the dining room fills with three-dimensional hologrammatic images from Hollywood films. Monsignor Frère cries with delight as the spaceship from *E.T.*, like some gorgeous illuminated Christmas tree, comes sliding down from the ceiling and from it advance tiny, child-like aliens. Mrs Cortland laughs with joy when Han Solo comes sauntering into the dining room as if it were an otherworldly cantina. 'How are you, Gorgeous?' he says, and winks at her.

Finally, the starship from *Close Encounters of the Third Kind* comes rumbling in for a landing. But the laughter stops when the hatch of the starship opens and Rudi Giger's monster from the *Alien* series lurches out, a towering insectoid with a long, tightly coiled tail. With a sudden pounce it is on Professor Van Straaten, nuzzling him with its phallic head and inserting its inseminating organ into his stomach like a stinger. In a moment the Professor's stomach bursts open and from it comes a small foetal alien which begins to devour him.

The holograms freeze and disappear. 'Quite a prelude,' the Professor tells Miss O'Hara as he mops his brow – and checks his stomach. 'It's completely put me off the rest of my dinner.'

Miss O'Hara laughs impishly. 'Now that I have your attention, let me focus on the large number of films and television series in the science-fiction genre that came out of Hollywood during the turn of the millennium. In them we see our wonderment at the universe."

'Is anybody out there?' Captain Craig muses.

'Space, the final frontier,' Dr Foley intones. 'These are the voyages of the Starship Enterprise. Its five-year mission to explore strange new worlds. To seek out new life and new civilisations. To boldly go where no man has gone before.'

When everyone laughs, Dr Foley hugs Miss O'Hara. 'See, darling? No need to apologise for not being a scientist. You hold all of us in your thrall.'

'So much of our fascination with other intelligent life in the universe is in those films,' Miss O'Hara continues. 'At their most benevolent, the extra-terrestrials were like Superman, who came from his doomed planet of Krypton with superhuman powers which he put to use helping our own. They were wish-fulfilment films in which the aliens came to preach anti-war messages or brought gifts of extra-sensory perception so that we could advance, emotionally and spiritually. Such films appealed to our own humanity, our yearning to become better people and to save our planet. When such aliens appeared on film they were often small, therefore not to be afraid of, or childlike and filled with light. They were like E.T. with that glowing finger of his, marooned on our planet –'

'E.T. phone home,' Monsignor Frère interpolates.

'The aliens were, in fact, not aliens at all but mirror-images of ourselves as we wanted to be. They were bipedal. They had faces and hands. They were, in all respects, just like us, but better.'

'They offered redemption,' Monsignor Frère says, his eyes lighting with pleasure. 'Often they featured a Messianic hero like Paul Atreides in *Dune* or the rebel prisoner, Riddick, in both *Pitch Black* and *The Chronicles of Riddick*. I have read a scientific paper which even critiques *The Terminator* from a Christian perspective. The cyborg terminator in the first film becomes the John the Baptist figure in the second film; he saves Sarah Connors, the Mary figure, so that she is able to give birth to her son, John Connors, the new Christ. And, of course, in *2001: A Space Odyssey*, man becomes a star child.'

'Don't get carried away,' Professor Van Straaten interrupts. 'Satan was also referenced in those films.'

'Behave, you two,' Mrs Cortland scolds them. 'Let Miss O'Hara continue with her dissertation. Go on, dear, and don't let those two men appropriate your thesis.'

'But Professor Van Straaten is right,' Miss O'Hara says. 'The interesting point about the films is that most of them favour a malignant vision of what intelligence is out there, and a more dystopian view of our future. For instance, there's the dark, apocalyptic world of Ridley Scott's *Blade Runner*, or of *A Clockwork Orange*, *Logan's Run*, *The Last Man on Earth*, *The Omega Man* and *Soylent Green*. All are built on intriguing premises, like *Planet of the Apes*, and most are filled with a paranoia, a fear that what waits in the universe are not angels but devils. They embody the fear of the unknown. Thus, the kind of aliens that were mainly portrayed were from the more fevered and nightmarish rooms of our imaginations. They were like *The Thing*, a real horror from beyond our solar system. The movie opens with the discovery of a flying saucer embedded in the ice at the Arctic Circle. Human rescuers recover the alien survivor and thaw it out but, like many of these aliens, this one then goes on a rampage, draining men of blood.'

'Yes,' the Monsignor agrees. 'Blood is sacramental in these movies.'

'Keep watching the skies,' Captain Craig nods. 'Those films fed Western paranoia. They were really about America's fear of the Soviet Union and, later, about the wars in the Middle East or against global terrorism. Their American-centric vision affirmed America's divine right to intervene in any war on the planet. But not even America could save us in the end.'

'And what was the purpose of the aliens?' Miss O'Hara continues. 'In films like *Mars Attacks* and *Independence Day* they are intent on world domination. In some cases, like in *The Quartermass Experiment* and *War of the Worlds*, they have been waiting for just this moment to strike. In *The Terror from Beyond Space* and *Earth Versus the Flying Saucers*, humanity enters into an archetypal life-or-death struggle with the aliens, who are always inhumanly strong and virtually invulnerable to conventional means of attack. In other films, like *Invaders From Mars*, *Invasion of the Body Snatchers*, *They Live*, *Supernova* and *The Puppetmasters*, they come to take over the intelligence of all humankind. They want us for their experiments, as in *The X-Files*, where they're supported by sinister government agencies, or they abduct us. In *Aliens* and *Predator* they are mindless creatures driven by the need to propagate for the survival of their own species.'

'In space nobody can hear you scream,' Professor Van Straaten remembers.

'Sometimes they are led by an alien queen,' Mrs Cortland contributes. 'I recall the chilling yet strangely erotic Borg queen in *Star Trek: Insurrection*, committed to enslaving human minds to the Borg collective. Though the most terrifying queen of all is the alien queen in *Aliens*. She's huge, monstrous, and when Ripley and her soldier forces blunder into her nest, she is hatching her eggs. As the eggs burst, the larvae feast on the succulent bodies of hapless human beings. The final confrontation between the alien queen and Ripley is really a working out of male fantasies of two women going head to head.'

'Nobody on Earth is safe from aliens,' Miss O'Hara continues, 'alive – or dead. There's a particularly sinister episode of *The Outer Limits*. It's called 'The Second Soul', and it's about a dying race which has come to Earth to save itself from extinction. They are a saprophytic race who can only survive on dead things, and governments therefore allow them to enter the corpses of the dead. Once this occurs, the corpses are reanimated. The episode ends with a child being born of the aliens and a dead corpse – a somewhat warped vision of the resurrection.'

Miss O'Hara tries to smile. 'Professor Van Straaten, nobody was safe from the scientists either! You were either mad, as in *The Island of Dr Moreau*, or psycho, as in *Doctor Strangelove*, or obsessed, as in *Demon Seed* and *The Island*, or crazed, as in *Forbidden Planet* – or all of the above. In *Aliens* you stupidly wanted to save the alien for scientific research even at the risk of human extinction.'

She gets up from the table and goes to the windows to look out at the blackness beyond. 'Carl Sagan thought we lived in a huge, demon-haunted world. Once it had been peopled by gargoyles and devils. Then it became populated by vampires, werewolves and zombies. And which side were the scientists on? Not ours, but theirs. With science fiction, our obsession with these dark visions strode out of the screens in new incarnations: the same, but new, and with greater presence and invincibility. All those space monsters did indeed represent the Devil, Professor. And do you know what? They were really only nightmare abstractions of ourselves. Our dual nature is reflected in the character of Anakin Skywalker in *Star Wars: The Revenge of the Sith*. When Anakin turns his back on the Jedi, he turns from Christ into Devil as Darth Vader. By doing so, he enshrines the belief that just as man cannot save himself, neither can he find any deliverance from the universe.'

Mrs Cortland leans over to Dr Foley. 'Go to her, Eliot,' she says.

And when he does so Miss O'Hara leans gratefully into his arms. 'The fact is,' she says, 'you have to be careful what you wish for. All of us have looked up at the night, wished upon a falling star and hoped there was somebody else out there. We've dreamed of great alien civilisations that could give us the chance to transform into beings of greater compassion and wisdom and not be like the nightmarish demons that fill our imaginations. We have hoped that they might bring us perfection, innocence and all those qualities of humanity we have always aspired to. Even if they existed, humans are such a toxic species that any aliens would have been wise to stay well clear of us. As it happens, in all our voyages through space, we have not found any other life at all. There are no aliens out here, hostile, friendly or otherwise. There's been nothing.'

Miss O'Hara turns to the others. 'So, my contribution to tonight's debate? The most transforming event in cosmological science? It's this: the discovery that man is truly alone. It has always, only and ever been in this huge immensity just – us. And if we couldn't find our salvation and transcendence from within ourselves, what hope was there ever that we would find it out here?'

Tears of hopelessness twinkle on Miss O'Hara's cheek.

If your heart is in your dreams,
No request is too extreme,
When you wish upon a star
Your dreams come true –

At that moment, Hemi interrupts the dinner.

'Captain, we are now in the final stages of our journey. The time has come to institute braking. May I begin inverse gravitational slingshot procedures?'

'Yes, of course, Hemi.'

Captain Craig and his dinner guests watch with excitement. What will the universe look like in the other extreme of Te Kore? As *Endeavour* slows, whirling around a giant black hole, a panorama of a jewel-box sky studded with glittering galaxies – wall after wall of them – is revealed.

The excitement turns to dread. 'Oh my Lord,' the Professor whispers.

The walls are advancing on the ship. They swell, metamorphose, overlap, combine and become super-dense.

Panicking, Miss O'Hara turns her head into Dr Foley's chest. 'Oh, hide me somewhere, Eliot!' But he knows there is no escape, nor anywhere to hide. The walls loom nearer.

Mrs Cortland begins to shiver. Uncontrollably. *Has it finally come to this? Just us? Are we the last to die?*

The others recognise the symptoms – they've all felt this way at one time. The only way to combat it is human touch – and Dr Foley puts his hand out to Mrs Cortland and grips hers. His warmth seeps into her.

'Peter would have been so proud of you,' he says to her.

Mrs Cortland gives a wry, sad nod. 'Yes,' she says. 'We've actually made it, haven't we –'

At that moment, Captain Craig sees Aunti-3 whizzing off like Tinkerbell. 'Who's afraid of the big bad wolf,' she sings, 'the big bad wolf, the big bad wolf? Who's afraid of the big bad wolf, oh no, no not I!'

Aunti-1 and Aunti-2 set off in hot pursuit. 'Wait up, girl, wait up!'

'Hemi, where are they going?' the Captain asks.

'We've received another message, sir. The aunties have gone to investigate.'

Very soon, the aunties report back. They are chittering and chattering excitedly. 'Are you ready for the incoming message?' Aunti-1 asks. 'Coming ready or not.'

The screen flickers with a mathematical equation.

PROPOSITION:
Set

$$(1) \quad |t = 0\rangle = \hat{a}^{\dagger}_{\alpha \wedge \Omega}|\text{vac}\rangle = \int d\xi \, |\xi\rangle_{\oplus}|\xi\rangle_{\star}.$$

'Sir, do you recognise it?' Hemi asks.

Captain Craig begins to scan the equation:

BEGINNING CREATED HEAVEN

And the meaning dawns on him.

'I think so, Hemi,' he answers. 'It's the first verse from the Book of Genesis.'

9.

'So this is how the universe ends,' Professor Van Straaten says.

'Or begins,' Monsignor Frère smiles.

The *Endeavour* is totally surrounded by gleaming walls of galaxies – millions and millions of them. It's a dizzying, disorienting sight as the walls continue their advance, towering around the ship. They are relentless and triumphant – and at their coming the voids are crushed from existence. Even Te Kore itself begins to collapse in the throes of death.

Aue, e Te Kore e,
Haere atu, haere, haere,
Haere –

*

Captain Craig recalls Captain Cook's second voyage to the southern seas. On 12 December 1772, Cook found himself at the edge of an endless pack of ice. On 17 January 1773, he had crossed the Antarctic Circle.

'Is my sense of wonderment,' Captain Craig muses, 'any different from Cook's when he reached the end of his world?'

He looks out the windows. Through the last remaining shards of Te Kore the ship sails, navigating the collapsing walls of the galaxies. All *Endeavour*'s engines are whining at maximum capacity, trying to maintain stabilisation. The light-wings are reefed, minimising her profile so that the ship is not overturned by the cosmic winds buffeting from all around, and Hemi has dropped seven space-anchors to keep her from drifting. Everywhere, the robotic engineering and service drones are trying to keep the shields operational against the gigantic maelstrom outside. Galaxies, suns and black holes continue to smash together.

Suddenly, there is no blackness. Instead, the view from the window flares with pure whiteness. The cosmic fluid is becoming so dense that sound waves start to propagate in it; they drown the sound of the ship's engines with the sounds of the entirety of creation.

Aue e te Ao Tawhito e,
Haere atu, haere, haere,
Haere –

The Artificial Intelligence is at full stretch. The power generators have moved into the red. The ship yaws, riding the currents of time as best it can.

'Nephew,' Aunti-2 scolds. 'Power down, man!'

'I'm sorry, Aunti-2,' Hemi answers, 'but I must maintain maximum stabilisation.' He has begun to use the light-wings in various configurations but they are whining with the stress load.

'Aunties,' Captain Craig asks, 'can you assist?'

'Sure thing, Captain. Why didn't you ask us earlier?'

They chatter to one another and then, agreed, whiz below *Endeavour*. There, they position themselves beneath the hull, cradling the vessel.

'Easy does it, old girl,' they say to the ship. 'Not long to wait now.'

The ship regains trim. The power generators move into green. The radiation shields hold. Even so, Captain Craig intercepts the fear on everyone's face. Mrs Cortland stares at him. She holds Miss O'Hara in her arms. She lifts her face so that the light shines full upon it.

Captain Craig walks towards her and joins her. 'Do you know the story of Galileo Galilei's first appearance before the Florence Academy?' he asks.

Mrs Cortland shakes her head: 'No.' She looks taken aback.

Monsignor Frère overhears. 'If I may, Captain? Galileo was a young man at the time – it was 1588 – and midway into his 20s. But he already had a reputation as a man of science, so the Academy asked him if he could solve a problem which men of the arts, men of letters, had been debating for some years. The debate centred on Dante's *Inferno* and on the dimensions of Hell and of Satan.'

Mrs Cortland sees Dr Foley and Professor Van Straaten standing at a distance. 'Come, Eliot, join us,' she says. 'You too, Professor. After all, we've all come this way together.'

'I know the story too,' the Professor says. 'It was one of the great debates between the sciences and the humanities. Galileo took up the challenge and, when he made his appearance before the Academy, he began by saying, "Dante's Hell is shaped like a cone one-twelfth the total mass of Earth. Its vortex is the home of Lucifer, who stands locked in ice halfway up his gigantic chest. His belly button forms the very centre of the Earth. From Lucifer, sectoral lines extend to Jerusalem on the Earth's surface and east to some unknown point. Inferno itself is a vast amphitheatre divided into eight levels. In the fifth level are the swamp called Styx and the wicked City of Dis. Here, the heretics suffer in the presence of Lucifer himself." You can imagine,' Professor Van Straaten chuckles, 'the reaction of the Academy to Galileo's calculations!'

'But there was more to come,' Dr Foley adds, nodding his head. 'Galileo went on to say that there was a relation between the size of Dante the man and the size of the giant, Nimrod, in the pit of Hell and, in turn, between Nimrod and the arm of Lucifer. Therefore, if the Academy knew Nimrod's size, it could deduce the size of Lucifer.'

'It was all there in *Inferno*,' the Monsignor continues. 'The divine Dante had himself written of Nimrod that his face was about as long and just as wide as St Peter's cone in Rome. "Thus," Galileo said, "his face will be five arm-lengths and a half and, since men are usually eight heads tall, the giant's face will be eight times as large. Therefore, Nimrod will be 44 arm-lengths tall."'

'I'm puzzled, Captain Craig,' Professor Van Straaten interrupts. 'Referencing Galileo at this juncture?'

'Oh, I know why he has done it,' the Monsignor smiles. 'Dante the man was to the giant as 3 is to 44. According to Galileo's calculations, "The relation of a giant to the arm of Lucifer is the same as the man is to the giant. The formula then must be 3 is to 44 as 44 is to X. Therefore the arm of Lucifer is 645 metres. Since the length of an arm is generally one-third the entire height, we can say that Lucifer's height will be some 2000 arm-lengths. This is the size of Lucifer."'

'The point of the story,' Captain Craig says, 'is that some historians say this was the inspiration for Galileo's further scientific investigations. If he could measure Hell and Satan, why not Heaven and God? And so he began his life-long assault on the stars. He thought Heaven had its own language. You had to understand its geometric figures before you could hear it talking. In particular, Heaven had a divine secret –'

'What was it?' Miss O'Hara asks.

'That the language of God is mathematics.'

Again, Hemi interrupts the conversation.

'Sir,' he says to Captain Craig, 'we've almost reached zero.'

Already? Captain Craig has a sudden urge to put the clock back – oh, back 1401 billion light-years, back, back, all the way back. Back to a place called Earth, a valley that he came from, a people he belonged to, a wife and two beloved infant children whom he loved. He would travel all that way just for one second, one second only, to embrace them. Would it be worth it? Yes. Oh, yes.

'The interesting question is which zero?' Professor Van Straaten observes. 'Have we gone forward to a new beginning or back to the beginning –'

'Either way,' the Monsignor prays, 'let's hope that we do a better job of it this time.'

'Captain Craig,' Hemi interjects. 'I have intercepted a countdown. Would you like it displayed for everyone to read? Thank you, Captain.'

All the screens show the following:

PROPOSITION:

Set

(1) $$|t = 0\rangle = \hat{a}^{\dagger}_{\alpha \wedge \Omega}|\text{vac}\rangle = \int d\xi \, |\xi\rangle_{\oplus}|\xi\rangle_{\star}.$$

With

(2a) $$\rho_{\oplus}(t) \equiv \text{tr}_{\star}(|t\rangle\langle t|)$$

undetermined; and

(2b) $$\hat{\boldsymbol{E}}(\boldsymbol{r})|t = 0\rangle = 0.$$

And evolution in time

(2c) $$|t\rangle = \hat{U}_{\alpha \wedge \Omega}(t, 0)|t = 0\rangle.$$

Let there be $t_l \in (0, t_1]$ such that

(3) $$\hat{\boldsymbol{E}}(\boldsymbol{r})|t_l\rangle \neq 0.$$

And $\hat{P}_{\alpha \wedge \Omega}|t_l\rangle = |+\rangle\langle +|$:

and there is an ω such that

(4a) $$\hat{\boldsymbol{E}}(0)|t\rangle \neq 0, \qquad 0 < [\omega t]_{2\pi} \leq \pi,$$

(4b) $$\hat{\boldsymbol{E}}(0)|t\rangle = 0, \qquad \pi < [\omega t]_{2\pi} < 2\pi.$$

Define

(5a) $$D \equiv \{t | 0 < [\omega t]_{2\pi} \leq \pi\},$$

(5b) $$N \equiv \{t | \pi < [\omega t]_{2\pi} \leq 2\pi\}.$$

And

(5c) $$\omega t_1 = 2\pi.$$

AND Let there be $t_H \in (t_1, t_2]$ and a partition $W_< \cup F \cup W_>$ of $\mathbb{R}_3$ such that

(6) $$\hat{N}_{\text{H}_2\text{O}}(x, y, z)|t_H\rangle = 0, \qquad (x, y, z) \in F.$$

and

(7) $$\int_F \hat{N}_{\text{H}_2\text{O}}(x, y, z)|t > t_H\rangle = 0.$$

Identify

(8a) $$\star \longleftrightarrow F.$$

And

(8b) $$\omega t_2 = 4\pi.$$

AND Let

(9a) $$W_< = W'_< \cup L,$$

with $W'_<$ connected, and let there be $t_E \in (t_2, t_3]$ such that

(9b) $$\hat{N}_{\mathrm{H_2O}}(x, y, z)|t > t_E\rangle = 0, \qquad (x, y, z) \in L.$$

Identify

(10a) $$\oplus \longleftrightarrow L;$$

and define

(10b) $$C \equiv W'_< :$$

And $\hat{P}_{\alpha \wedge \Omega}|t_E\rangle = |+\rangle\langle+|.$

And Let there be $t_p \in (t_E, t_3]$ such that

(11) $$\oplus \Rightarrow \hat{N}_{S(\mathrm{A,G,C,T})}|t_p\rangle \neq 0, \qquad S(\mathrm{A, G, C, T}) \in P_{LT},$$

And for $t > t_p$

(12) $$\langle \dot{\hat{N}}_{S(\mathrm{A,G,C,T})} \rangle > 0.$$

And $\hat{P}_{\alpha \wedge \Omega}|t\rangle = |+\rangle\langle+|.$

And

(13) $$\omega t_3 = 6\pi.$$

AND Let there be $t_L \in (t_3, t_4]$ such that

(14a) $$\left. \begin{array}{l} \exists\, \{\boldsymbol{r}_D\} \subset F \text{ s.t. } \hat{\boldsymbol{E}}(\boldsymbol{r}_D)|t > t_L\rangle \neq 0 \\[6pt] \exists\, \{\boldsymbol{r}_N\} \subset F \text{ s.t. } \hat{\boldsymbol{E}}(\boldsymbol{r}_N)|t > t_L\rangle \neq 0 \end{array} \right\} \longleftrightarrow D \cap N = \emptyset;$$

and let

(14b) $$\{\boldsymbol{r}_D\} \wedge \{\boldsymbol{r}_N\} \Rightarrow \pm \wedge [\omega t]_{2\pi} \wedge [\omega t]_{182.5\pi} \wedge [\omega t]_{730\pi} :$$

And let

(15) $\{r_D\} \wedge \{r_N\} \subset F \longleftrightarrow \star \Rightarrow \hat{E}(r_\oplus)|t > t_L\rangle \neq 0.$

And

(16a) $\hat{U}_{\alpha \wedge \Omega}(t > t_L, 0)|\text{vac}\rangle$ s.t. $\exists \odot \equiv \{r_D\} \wedge \big(\equiv \{r_N\}$

with $\odot > \big($ and

(16b) $\odot \Rightarrow \hat{E}(r_\oplus)|t > t_L\rangle \neq 0, \qquad 0 < [\omega t]_{2\pi} \leq \pi,$

(16c) $\big(\Rightarrow \hat{E}(r_\oplus)|t > t_L\rangle \neq 0, \qquad \pi < [\omega t]_{2\pi} < 2\pi.$

And

(17) $\odot \wedge \big(\in F \longleftrightarrow \star$ s.t. $\hat{E}(r_\oplus)|t > t_L\rangle \neq 0,$

and

(18) $\odot \wedge \big(/ D \wedge N \longleftrightarrow D \cap N = \emptyset :$

and $\hat{P}_{\alpha \wedge \Omega}|t\rangle = |+\rangle\langle+|.$

And

(19) $$\omega t_4 = 8\pi.$$

And Let there be $t_f \in (t_4, t_5]$ such that

(20a) $W'_< \Rightarrow \hat{N}_{S(A,G,C,T)}|t_f\rangle \neq 0, \qquad S(A, G, C, T) \in F_{SH},$

and

(20b) $W'_< \Rightarrow \hat{N}_{S(A,G,C,T)}|t_f\rangle \neq 0, \qquad S(A, G, C, T) \in F_{WL}.$

And for $t > t_f$

(21) $$\langle \hat{\dot{N}}_{S(A,G,C,T)}\rangle > 0 :$$

and $\hat{P}_{\alpha \wedge \Omega}|t\rangle = |+\rangle\langle+|.$

And let

(22) $$F_{SH} \otimes \in W'_< \wedge F_{WL} \otimes \in \oplus.$$

And

(23) $$\omega t_5 = 10\pi.$$

AND Let there be $t_b \in (t_5, t_6]$ such that

(24) $\oplus \Rightarrow \hat{N}_{S(A,G,C,T)}|t_p\rangle \neq 0, \qquad S(A, G, C, T) \in B_{ST},$

And for $t > t_b$

(25) $\langle \dot{\hat{N}}_{S(A,G,C,T)} \rangle > 0.$

and $\hat{P}_{\alpha \wedge \Omega} |t\rangle = |+\rangle\langle+|.$

AND Let there be $t_{\text{⊕}} \in (t_b, t_6]$ such that

(26a) $\exists\, \text{⊕}(A, C, G, T)$ s.t. $\text{⊕}(A, C, G, T) \models \alpha \wedge \Omega :$

and let

(26b) $\text{⊕}(A, C, G, T) > S(A, C, G, T)$

$\forall\, S(A, C, G, T) \in F_{SH} \cup F_{WL} \cup B_{ST}.$
So

(27) $\hat{a}^{\dagger}_{\alpha \wedge \Omega} |\text{vac}\rangle \longrightarrow \text{⊕} \equiv \text{O} \cup \text{Q} \models \alpha \wedge \Omega.$

And let

(28) $\text{⊕} \otimes$ s.t. $\text{⊕} / \oplus \wedge \text{⊕} > S\ \forall\, S \in F_{SH} \cup F_{WL} \cup B_{ST}.$

AND

(29) $\forall\, \text{O} \vee \text{Q} \in \text{⊕}\ \exists\, P_{LT}$ s.t. $\text{O} \cup \text{Q} \Rightarrow \text{O} \cup \text{Q}.$

And $\forall\, S(A, C, G, T) \in B_{ST} \cup F_{WL}$ there exists P_{LT} such that

(30) $\text{O}_S \cup \text{Q}_S \Rightarrow \text{O}_S \cup \text{Q}_S :$

Q.E.D..
And

(31a) $\hat{P}_{\alpha \wedge \Omega} |t\rangle = |++\rangle\langle++|.$

And

(31b) $\omega t_6 = 12\pi.$

The mathematical formulae stop.

Mrs Cortland presses Miss O'Hara's hands reassuringly. 'We'll be all right, my dear.'

The *Endeavour* rocks. The three aunties are like lifeboats beside it. They are chattering as if they know what is going on.

'Here we go, girls,' Aunti-1 says.

'Captain,' Hemi says, 'I'm getting readings of huge energy forces coalescing all around us. Something is happening.'

The ship rocks, and rocks again. Then everything is happening at once. Echoes of the whole titanic history of the universe are crowding in, pushing and jostling with each other, fighting for the last sliver of space. Every movement ever made. Every word ever spoken. Every television show ever broadcast. Every ray of starlight ever shone. Space is collapsing under its own weight.

'It won't be long before the gravitational waves overwhelm us,' Captain Craig says. In his mind's eye he sees a trillion black holes in collision. Tears spring to his eyes at the thought of all that human history – whatever the self-destructiveness, there had also been so much *life,* so much hope, so many dreams achieved and triumphs witnessed. As one of the last survivors, Captain Craig raises his voice in poroporoaki, in defiant tribute to the generations upon generations of men and women who had lived in this life and world:

Tena koutou nga iwi katoa o te Ao,
Te Huinga o te Kahurangi,
Tena koutou –

'Look!' Professor Van Straaten says.

His voice is hushed, tinged with awe. The universe may be in its death throes, but a countervailing option is making itself apparent on all Hemi's screens. It's an image that is so familiar, oh so familiar.

'The double helix,' Dr Foley whispers.

He rushes to the window to look out, closely followed by the others. The double helix is a million miles high. It is also the koru pattern, twisting and turning – a dazzling signal in the flaring, blossoming light. The spiralling helices flow around the *Endeavour,* enclosing it, locking it into the world of *now.* The world of *us.*

'What is happening?' Mrs Cortland asks.

'We're getting a second chance,' Dr Foley says.

There is a moment's silence. Then Hemi's computers go haywire. Huge twisting ribbons of energy, of heating and expansion and contraction, come pouring around *Endeavour.* The ship becomes the binding central nucleus, sealed into the double helix. Captain Craig looks at Professor Van Straaten and nods to him:

'You are right, Professor. The end of the universe is not a place. It is a time. It's alpha and omega, the beginning and the end.'

Turbulent and 10-dimensional, time begins to vibrate in different modes, splitting, rejoining and moving fluidly backwards and forwards – and *Endeavour* rides upon time's waves through an ever-changing, swirling continuum.

Hemi's voice comes, soothing, to Captain Craig. 'Ten seconds to the first impact from the universe's shock waves. Nine, eight, seven . . . Sir, time to open the codicil to your secret instructions.'

'Open,' Captain Craig orders. On all screens there appears one word:

RESET.

'Four, three, two, one –'
'Do as ordered,' Captain Craig says.
'Zero.'

10.

Time fades to nothing.

Real nothing. Not even space. Outside? Outside there is no outside.

Space is time and time is space. Space enough, perhaps?

All there is, is the double helix, floating within an immense womb of blackness. At its centre, *Endeavour*, cradled by the three aunties.

'Up, up and away!' Aunti-3 sings. 'Let's fly up and away like a beautiful balloon!' Her voice is ecstatic. 'Yes,' Aunti-1 and Aunti-2 respond. 'We've done our job.'

Vibrating, singing, pulsing, the double helix splits, streaks, fragments, rejoins, coalesces, mottles, blossoms and flares. At each transformation, it refracts and flashes like a scintillating crystal ribbon, repeating itself over and over, twisting and turning, binding together futures without end:

Now is everything and everything is now.

Breathtakingly beautiful, the complex helices twist into the primeval Word:

The Greek letter Ψ.

The wave function of the universe, the Word reveals all the infinite, heart-aching wonder of all our possibilities.

Then:

$$H \, |\Psi> = 0$$

The ship is like a bloodied jawbone thrown through the air. Time begins again. The energy is more than 10 billion billion billion nuclear detonations. It roars over the *Endeavour*. In a trice, the ship is incinerated. Her avatar, Hemi, gives a huge, deep sigh. The glorious aunties are like angels on fire, fluttering into oblivion.

Captain Craig feels an intense pain. He looks at Mrs Cortland, Miss O'Hara, Monsignor Frère, Professor Van Straaten and Dr Foley, wanting to reassure them. But of what? Just before the endorphins kick in, he has a regretful thought:

'But we didn't have time to say goodbye to each other.'

And he is falling.

He feels a dizzying rush of acceleration, as if he is being sucked at headlong speed down a tunnel of dazzling light. Onward and onward he roars, and the sensation is so delirious that he wants to laugh and laugh:

Tuia i runga, Tuia i raro,
Tuia i roto, Tuia i waho,
Tuia! Tuia! Tuia –

All of a sudden he is through the tunnel and suspended above the blackness, watching the primordial fireball and the way its wave of light is moving so fast, creating a new cosmos. He becomes frightened and closes his eyes.

And he is falling.

When he opens his eyes he sees a wild landscape in the country of his birth, New Zealand. It is so familiar to him that he laughs with relief. Suddenly a mist descends and he is lost in it.

This has happened before, he thinks. The mist opens and he sees three elderly women from the village walking in front of him. He runs after them. One of them turns and asks, 'Went the day well, sir?' The second says, 'Not you again.' And the third says, 'You're always losing directions. Well, you're almost at your destination, lad. There it is.' She points to a faraway farmhouse on the other side of the valley. 'You had better make haste,' the old women say. 'Night is coming and, with it, a fierce storm.'

He walks to the farmhouse. A light is coming from the window and, framed within the light, someone is watching him. As he approaches, the door opens. Mrs Cortland is there.

'Thank goodness you were able to make it home before dark,' she says. 'Come in, come in.'

He enters the house and sees Monsignor Frère, Dr Foley, Miss O'Hara and Professor Van Straaten are having a cup of tea and chatting. 'Let the Captain have a place by the fire,' Mrs Cortland scolds. 'He'll catch his death. There's room for one more.'

The five people in the room smile at Captain Craig as if they have known him all their lives. Monsignor Frère comes to join him. 'Don't be afraid,' he says. 'There are some questions that science cannot answer. They may know what, how and when, but while faith and reason have co-existed in scientists as notable as Isaac Newton and Albert Einstein, only theologians know *why*.'

The Captain shifts on his feet nervously. Mrs Cortland brings him a cup of hot tea. 'This will bring you back to life,' she says.

A few moments later, Professor Van Straaten comes to talk to him. 'You shouldn't believe everything you see,' he says. 'At the end of the universe things still go bump in the night and every grave, opening wide, lets forth its sprites and demons. But Fate is always kind. She gives to those who love their secret longing.'

'Do I know you?' the Captain asks the Professor. 'Do I know any of you?'

'Of course you do!' Professor Van Straaten laughs. But he is uncertain, and his laughter fades away into bewilderment. 'Because if you don't, then who are we?'

Monsignor Frère winks at him. 'Interesting, isn't it?'

Mrs Cortland claps her hands for attention. 'It's time to go,' she says. 'Dr Foley and Miss O'Hara, are you ready? The next great adventure is about to begin. As for the Captain, he has his own journey. Goodbye, my dear.'

*

And he is falling.

Captain Craig finds himself in a small white room. He is alone. He is naked.

How did he get here? Who is he? Why is he here? He begins to scream and pound on the walls. He falls to the floor in a foetal position, howling, hugging himself and weeping.

Time is limitless. How long has he been here? He does not know. He sleeps, and time continues to go by. Suddenly he hears a voice: *Open your eyes*. Perhaps it is his own voice; he doesn't know.

When he does wake up, he sees that there is a wardrobe. In the wardrobe is a white suit. His heart is beating loudly. He dresses in the suit.

A door appears in the room. Above the door is a clock. The time is just before midnight.

He finishes dressing. He sees a mirror on the wall. He inspects his appearance. Combs his hair. Smooths out his trousers.

Takes a deep breath. Walks to the door.

Turns the doorknob.

Opens the door.

ELSEWHEN

THAT GREAT SCRAPYARD WHERE TIME IS PARKED AND
WHERE THE USUAL COMMON SENSE IS NO HELP AT ALL

LLOYD JONES

Etiquette

No one with any sense of sophistication ever says, 'Remember the
time . . .'
Instead you might ask yourself this: with the free fall of night
the uprooted moment
what do we need graves for?

Elsewhen's heroes and adventurers

Calvino for daring us to believe it was possible to ascend a stepladder
to the moon.
R.L. Stevenson for his map of an island that did not exist (Treasure
Island existed in our imagination long before satellites in the 1980s
picked up the presence of hitherto unknown islands in the Canadian
Arctic).
Borges for his immense libraries of potentiality.
Alexander Dalrymple for his belief in something (Terra Australis) at
a time when it did not exist on any map.
Samuel Butler for turning nowhere into somewhere.
Demeritus, the Sicilian monk, who slept under the sky for 48 of
his 55 years and was convinced that space was comprised of dots.
He postulated that these grains in the sky could be numbered and
joined in any old sequence.
Eratosthenes for daring to measure space greater than the human
eye could see.
Gödel for shining his lamp on eternity and for so helpfully showing
the way forward: 'Only fables present the world as it should be and
as if it had meaning.'

Two cautionary tales
Though just one should suffice

Jenny's folly

Jenny dropped a timepiece down a deep well. She obsessed over the loss and spent the rest of her days and nights trying to fish up the timepiece, right up until
the day
she died.

Silly old Jenny.

And then there is the ongoing tale of 'Ne Plus Ultra' (nothing lies further). In this way the Pillars of Hercules marked the end of the known world.

Exploratory notes

I knew about language long before I lifted my eyes off the page to the unruled night sky, which to my untrained eye lacked form. I could not see all those animals and shapes that are said to lurk up there. But I understood the space between the stars, because the language I like best is that which is left unsaid, and when the professor told me the latest news from the cosmos, courtesy of a conference in smoggy Beijing, of all places, that the reality we see may only be the gaps between, I understood immediately what he was on about.

It's the gaps we like, that draw us. The bits we can't see. The rest is just clutter.
I had never before heard the name Eratosthenes, which now seems unforgivable. I never knew that time could bend like sheet metal. I sort of accepted that time came packaged in clocks and watches. I never realised there was such a thing as big time and little time. Little time belongs to us. It sits on our shoulder from the time we are born and rides us all the way to the grave. Big time belongs to the cosmos. Big time is showtime – space is a fat boy who just gets fatter. I suppose I was more used to hearing about *the flow of time* bubbling and swelling into the caverns of the imagination and myth.

I never knew about the place of ego in time. Or that the eyes and ears of our record-keepers are attuned to time that makes the biggest noise. The moon landing, Hiroshima, this and that assassination – these days are indelibly marked, set aside, retired in the manner of a number off the back of a famous athlete. 9/11. On that day, some parallel story whose noise had barely been heard up until this moment careened into our ear space. And the noise was stupendous. This is time with an ego. Yet we can be sure that somewhere else in the world, at that very moment of inferno, well-wishers were cheering and singing 'For she's a jolly good fellow' as the candles on the birthday cake fluttered out. Little time is inscrutably modest.

I suppose I was vaguely familiar with the popular 'garden swing' school of time. It is simple enough to grasp. As you swing through the air, back and forth, back and forth, there is no difference between this moment and the last or the future.

Bus timetables understand this. As a matter of necessity the past and the future are the same thing. That's why we so confidently line up at the bus stop without a bus in sight. A timetable is like a diver's guide rope. You need to know which way is up and where *where* is.

Six months have passed since I packed into a university theatre to attend a series of public lectures on new discoveries from the subatomic world. The news by the way is neither good nor bad. In the subatomic world there are no armies of darkness, or for that matter crusaders standing on a sun-lit rise. There just is – and a lot of it, waiting, spinning, every moment in possession of endless potentiality right up to the moment we set our eyes on it, and then everything changes. To be observed and to observe are not the independent states we thought they were. It seems a glance can change everything. The young woman blushes. The dog looks away. The would-be shoplifter puts the chocolate bar back on the shelf. Or take the woman in the Mavis Gallant story: during a Paris street riot she has bent to find a rock to hurl. Now, as she rises with her arm cocked ready to throw, her glance is intercepted by the disapproving frown of her future husband, a future magistrate, a man who until that moment was a complete stranger. And what does she do? She places the rock down. One action is prevented, another is set in train.

I suppose that is what happened to me. Though the transporting moment wasn't conveyed by a glance as much as by a flash of light.

With the use of light cones the professor was able to demonstrate the known regions of time; the present moment was represented by a flash which spread back and forwards. The professor referred to a diagram and I will do the same: think here of two inverted triangles which represent the transmission of light – light being the tool physicists rely on to plot the movement of time. The professor indicated present, past and future, and with what I remember as a grin tucked into the side of his cheek he identified 'elsewhen' out to the sides. *Elsewhen.* The moment I heard it spoken I accepted that I was hearing a place name. I saw events sailing serenely in the ether. Sail boats moving at the speed of light. Sail boats beached on the edge of shrunken seas. The old order lay about like tossed-up kelp from the last storm. I saw trees simultaneously lose their leaves and come into colour. I saw a great entanglement of time and event, something like one of those tip faces we have all stood before trying to make sense of so many things discarded. By now the professor had moved on to the real subject of the lecture. But I didn't hear. I was no longer listening because I was stuck in 'elsewhen'. I was stuck in this place which in all truthfulness the professor had alluded to in a whimsical way, to judge by the titter of the audience. Yet I thought I knew this place. I knew it foremost as a territory of the mind. I knew it to exist just as I had known there was a bottom to the sea before my little sister surfaced with a fistful of grey mud as proof of her reaching that spoken of and yet unreached destination. I remember the grey mud oozing between her fingers, the triumph of her gappy smile and lank wet hair, her absolute delight – that was mine too, years later, sitting among strangers in the lecture theatre. Absolute delight.

For a few days after the lecture I sat behind my desk at the hotel pondering the tools I might need to get me there. I enjoyed the conceit. After all, the great voyagers didn't leave port without a sextant to guide them. Varnished and beautifully illustrated globes rocked in creaky cabins. They loaded up on rum. They brought a change of clothing for all seasons. They brought with them a knowledge of algebra and a huge respect for Eratosthenes, the first man to measure a space too large to be seen with the human eye. But the parrot chirping into my ear was Kurt Gödel. It was this little man, more sparrow than parrot, who whispered the way forward. I could, if I wished, plot the coordinates of *Elsewhen* – more exactly than our friend, the professor, had managed with his vague map – through tales and fables. I thought if I looked hard enough I would find this place in the everyday transactions of life.

Tales from the kingdom of entangled states

The night clerk explains to an insomniac
the whereabouts of Elsewhen

Example one: Suddenly the lights come on in the picture theatre, and there is the audience – startled, annoyed, a little embarrassed. Lovers adrift in one another's arms. A tiny upright man in a heavy overcoat who knows how the scene with the Indians will end, having seen it all before. The lights go off. The audience sinks into darkness. The spiders creep back out of hiding. The Indians go on whooping and the engine driver is as we left him, sweating and hollerin'; while his assistant shovels in coal as fast as he is able as the train approaches the bend in the hill.

Example two: Cup Final Day. An immense roar goes up from the crowd as you pee all alone in the urinal at the back of the stand.

Example three: All the animals are gathered in the dining room of the manor. They gaze up at their mounted heads. They tread lightly over their skins pressed into ornamental floor rugs. They watch the diners around the table tuck hungrily into their thighs and shoulders. The deer that was too big to enter the door looks in from the window. It is snowing outside. A fire is blazing away, and above the hearth is a photograph of the lord of the manor kneeling beside the deer he shot which is also the deer looking in from the window, watching these good people tuck into their meal. There is his own flesh sizzling in a hot pan, which reminds the deer of his own hunger.

Example four: The American bombs are falling on Nagasaki. The lovers kneel to face one another. They are pledging themselves to one another for all eternity. There is a flash in the window. A soldier bends forward to inspect two shadows 'printed' against the wall. 'Hey Hank, come and take a look at this. What do these shadows remind you of?' The soldiers are dead. Their throats slit. They lie in the rubble. They share a cigarette and talk about what those shadows remind them each of: a woman by the name of Marlene, another woman by the name of Betty. Warm thighs. Great tits. They talk about these things. Look at them. The empty eye sockets of their skulls turned up to the sky. Skeletal hands at their sides, parade ground protocol to the end. On Marlene's mantelpiece is a photo taken by her dead marine boyfriend. She tells her grandchildren, 'This is the man I might have married.' Someone who actually thought to take a photo of such a thing. The thought passes. No. She could never have married Hank. And then he went and got his throat cut anyway. She tells her grandchildren: see how the people leave their shadows behind. Hank seems to think one in particular is the man. 'Marlene would have something to say about that,' Hank says, pointing.

The lie of the land

In the bay of recurring dreams you cast your line and reel in whatever dream appeals. You can throw back the ones you don't want, or dreams you thought you did only for them to turn into nightmares. These are the dreams you see drying out in the sun above the high tide line. Sometimes you see spiteful, unsophisticated types breaking up certain problematic dreams into small pieces which they bury in scattered places the length of the beach. But mostly the bay of recurring dreams is a happy place. Of course, it has become more predictable since it was bought up by the Americans and franchised out.

We liked to wade out in the tide up to our knees to see the bloated forms of film stars. One summer we challenged one another to wade out a bit deeper and touch the short fat guy who played Al Capone. I took up the bet; the anticipated hail of lead failed to eventuate. Al Capone turned out to be pink and surprisingly feminine, actually. But, through the transparent walls of the giant jellyfish's pink belly, it was possible to see the unmistakable dark shape of a rotund bald man, gun in hand.

Again, before the Americans came, as kids we would hide in the sand dunes and spy on other people's dreams. We were really looking out for naked people. We always looked out for pimply Ralph Scott, whose dreams never disappointed.

In the next bay round (which the Americans didn't know about at the time of their purchase), an ancient pohutukawa sends its willowy limbs out over the sea. On the outgoing tide you can tiptoe to the very end of the branches and dive off these boughs into a passing dream of your choice. The Americans don't suspect a thing. But only on the outgoing tide.

Visiting days

My mother holds her old, blind face up, smiling away – like she knows this is something she is expected to do. All around her she is surrounded by unsmiling assassins. The walls will not help her. The carpet, which is so carefully and lovingly vacuumed every day, will not lift a hand to help her. Out the window the sea laps the shore, just like a moment ago, just like yesterday and the day before, just like tomorrow and the day after. The sea laps the shore, the carpet and walls contain their treachery. My blind mum holds her blind face up. Is that a car she heard pull up in the drive? Perhaps it is a visitor. Down in the carport I see my sister get out of her red car with the roses that my mum cannot see or smell. I tell her who it is. I have to shout, which means she has to lean her ear towards my mouth, which means the smile must leave her wrinkled cheeks in order for her to hear. On the table next to her armchair is a photograph taken in another century. It is of my mum aged seven, and there she also has her ear cocked, as if she is listening out for the future. My sister comes into the room. The walls and carpet are slightly scared of her. They move back for her expensive leather smell and rich perfume. She pecks my mum on the cheek. My mum is smiling. She has a question. She actually spoke. We move in around her. We crowd the space like we did around the radio the day news came through that man had walked on the moon. She asks, 'How old am I?'

Art's accidents

That year there was a terrible crash. A car tangled up in the front of a truck, its front wheels lifted clean off the road – like a dog trying to squeeze through a cat flap. After the sirens faded, after the broken glass was swept and the blood hosed from the road, after the iron wrecks were towed, normal traffic resumed, and a day later when she stood on the side of the road she could no longer be sure where the crash had occurred. Every mark had been removed.

But had she followed the trail this is what she would have found. The wrecks were towed to a wrecker's yard and there the wrecks were crunched down to the size of two suitcases. They were dropped off to a metal sculptor who as a sideline makes garden ornaments, and two metal suitcases, once a truck and a car, were turned into a cat and dog which today stand playfully together on a front lawn outside a tarot reader's house across the bay.

Smile

Of all the seagulls I have ever seen – once, I happened to look up as a gull flew at a great height across my father's open grave, another time I saw one riding on top of a baby's carriage, another time there was one sightseeing from the back of a jersey cow, once a flock of them spread like an RAF squadron on the way to the tip, and another time a smart one stood on the balustrade eyeing the fish remains on the plates at The Fisherman's Table, and each time the seagull moved towards the fish remains its shadow spooked the several thousand herring that had come into the shallows of Oriental Bay for the late afternoon – this one is the most memorable. Click.

Time stops, then kicks on

A man stretches a chicken neck and brings down the axe, severing head from quivering body. The head is tossed into a bucket filled with chicken heads with their smart red Indian-chief headpieces. None look surprised or the slightest bit offended.

Another bird story

In drawer 4B in a temperature-controlled room beneath the national museum a number of views lie in the collection. There are too many to list here. But I will mention: the man and the woman lying on deck chairs beside the garden sprinkler which their naked four-year-old runs in and out of; I will mention the rainbow that certain winged insects foolishly follow, mistaking the trail for one of the ancient ones. I will mention the dog lifting its snout from its paws, its ears lengthening. I will mention the old woman stalled between clothesline and house (she has just heard her baby crying, now how could that be? When she last saw Brian he was fat and 42 years old, bulging inside a white singlet, smiling with a broken tooth that he never satisfactorily explained away), and I will mention the young man sitting on the lawn with a note from his girlfriend. These views are contained within waxeye 249. Views of the house fire and a dirty backyard story can be obtained from waxeye 250. The waxeyes appear to lie in pairs. Then, after a while, you think that maybe they are lying in fours, or even in a flock, before returning to the idea that each one in fact is an individual, and there is no point in reshuffling the dead birds to provide a more coherent view. These things happened, and in no particular order. But you could if you wished cut open bird 250 and pinch the view of the house fire and stuff it inside bird 249 to go with the last view of the disenchanted young man sitting on the front lawn, but only if you believe in conspiracy theories.

What the rock knows

There is a girl sitting on my lap.
There is a man sitting on my lap.
There is an old woman sitting on my lap.
There is a young woman sitting on my lap.
And in between times there is something the girl, the man, the old woman and the young woman call the lap of time.

And suddenly

Cathedral bells
The empty battlefield
The old man sits up in bed.

Upon leaving the hospital

Rush hour traffic at the roundabout
My mother has lost all feeling
No one in the photographs knows what is going on.

Look

In the window pane
The purple agapanthus
And the foaming surf

and

In the shop window
A trailing cloud
A naked mannikin.

And, it also happens . . .

Two things will have nothing to do with each other. For example, those gannets turning on a wing and crashing into the sea, and me lying on my bed looking out the window and thinking of my old mum croaking up at the nurse with her pills, 'Must I?'

A whale of a time

There are balls of time. There are great drifts of time that blow in and settle. There are unsteady pillars that fetch up, bending and wobbling. There are cracks in the universe through which time slithers and spills without pause or fuss. There are strands of time that fuse into great spaghetti junctions and throw up teasing moments of déjà vu. There are the giant swamps where time oozes and sets in its own mess. There are terrible wastelands where time hangs off row after row of meat hooks so that the future may look back in horror and know right from wrong. There are also giant plug holes where time is the biggest vanishing act in town. And there is something called eternity, once promised to me and Bruce Howell and Jim Shepherd by a man in a starched white collar. The man's face was an ancient parchment of severe lines and containment. Today we'd probably say he was an alcoholic. Anyway, all we had to do, he said, was follow him into the back of the church hall where orange juice and a plate of biscuits awaited. That was the first step to eternity. Bruce and Jim waited for the time it took for a lost car to pass on the road, and took up the offer. I preferred the butcher shop with its pleasing smells of sawdust and meat, an aroma which in later years I associated with the most anodyne of the senses – a way of making the unpalatable almost acceptable. Like war and famine on the telly. And always, as I remember it, the butcher slapping his mutton-chop arms down on the counter and breathing blood and sawdust over me, to ask, 'What do you know, Snow?'

What the fence knows

1942. Blistering heat. Cicadas. A child's sticky fingerprints.

1944. The tail of a kite.
Conversation overheard: 'Lovely day. Radishes looking good.'

1959. Conversation overheard: 'Lovely day.'
Cicadas.
'Cabbages looking good.'
'Don't. I don't want to get pregnant.'
'Keep your voice down. Someone will hear.'

1963. Blistering heat. Sticky fingerprints of a handicapped boy from down the street feeding bits of the fence into his mouth. Someone has told him the wood is really huhu bug . . . which his aunt calls 'brain food for a rainy day'.
What the fence knows is the sun and the rain and the wind, and

in between times, mistaken belief and complaint. The fence does not complain. It collects. It goes on collecting. Occasionally a word like *disputatious* drifts into range and the shadows of men gather to debate a boundary line.

There was a time (shall we call it) when the fence was also a tree.

It is a tree standing on a point overlooking the sea which appears to go one way, then the other. It is a tree that knows it is pointless to count the ways.

It is a seed falling through the air.

It is an idea.

What the fence knows
The orphan knows
As somewhere else.

Well, this is a common enough occurrence

She arrived at the door out of the blue. Said she'd lived here 35 years ago, here in this same house, and would we mind if she took a look around. Excuse the mess, we said – repeatedly. She did – but she did not necessarily forgive. She poked her head into all the rooms. It was weird. She knew where to go, which room was where, and nothing really seemed to come as a surprise. She stood in the door of the room which overlooks the back garden, once a beautiful garden, we heard, now overgrown (more apologies, which she ignored). The woman said, 'This is where four of my children were conceived. Only the bed-head used to face the window. We'd keep the window open and listen to the tuis. Look, there's one now.' And there was. A fucking great tui. The woman said, 'I wonder if it's Freddy?' And the woman began to chirp. The tui sang back. The two of them chirping until the woman broke off. 'Nah,' she said, 'it isn't him.' And to the tui – 'Nice try, sunshine.'

And

From Berlin, on TV, comes a report of citizens running smack into invisible walls. One woman is shown crossing the road. Two caregivers take an arm each. Her face is raised, eyes shut, mouth stretched wide with protest. She is sure she will be shot. They are agents. She is innocent. They? Well, those two on either side. They are escorting her to her death. She will become one of those lifeless bodies caught on the wire. She is. They are. Between history's dark negatives. The road as always.

A glass-blower writes from his hotel room

We wait in our vestibules – the ant, the small boy, the bird packed inside its egg. We wait to be unleashed with whatever wind is blowing that day, rain or shine. The tree awaits the bird; if not that one, then another, just like the last, as well as the one in the future. But for the moment we wait with our heads bowed and strapped into our parachutes. My father is calling up to me in the tree. 'Jump, Bryn. Jump for Christ's sake. Go on, jump. I'll catch you.' A soap bubble, trembling and bright, rises past my window. I get up from my sickbed in time to see the bubble rise past the peeling gutter with the rusted holes, past the proud chimney, past the danger. A bird flies on with a worm in its beak. I jump and my father catches me as he promised. Whenever I blow my glass I think of that soap bubble and my father's entreaty to me to just shut my eyes and trust. That's what the mountain of sand believes in; for its piety one day that mountain of sand will have the transparency of glass. My wish is to be buried in a glass coffin. I don't want to miss anything. And I want to be buried strapped into my parachute, just in case I should rise. I hope to rise before I am found out. For then I will drop like a stone, that's the moment I want my parachute for. But for the moment I am just dawdling. The night shows no sign of ending, and the drinks in the hotel bar are far too expensive.

Another day

My old mum has no idea it is her birthday today, but she will take it, as sportsmen say of victorious moments that could have and perhaps should have gone the other way.

How they met

She was getting off a bus and he was taking five from a film set. She was the one in motion, slapping down her skirt billowing in the wind. He was leaning against the bus shelter, casual, cigarette in hand. He stayed in character right up until he stepped away from the bus shelter to ask the good-looking woman for directions. Later he played Hook in *Hook*. She loved Robin Williams in Spielberg's movie of the same name, another Hook, spinning round and round in celluloid, on stage and off-Broadway, a retirement home of Captain Hooks roaming the hallways at night, tucking into the scones, some more able than others, some requiring sedation at night, some in baby diapers, a retirement home of Hooks waited on by a staff of Wendys. There she is getting up from her seat, sitting down again, sitting with her mouth open, her heart pumping, oh the relief, Hook is dead, the crocodile got him.

When people ask me I say

My father worked with steel. His hands were tough as. I saw him pick up gorse bare-handed once and throw it into a pile. Once I saw him cast his rod – we were after kahawai – and when the lure with its jagged hooks caught and embedded in his cheek he ripped those hooks out. Jesus, he didn't give it a second thought. He looked like a man who'd just missed winning Lotto by one number. Rage – pure rage it was. He stood on the end of the wharf with blood pouring from the wound in his cheek.

If he was alive today I'd have to hide my prissy clerical hands from his roving eye. Sooner or later, at the dinner table, there would be no escape. I'd be tricked into passing the salt, then he'd pounce, grab my hand and turn it over for all to see. And we'd enter that favourite pantomime of his.

'What do you call this then?'

'That's the letter S.'

'And that?

'That's O.'

'And that?'

'That's F.'

'I know where this is heading. I thought so. Might as well pass me the pepper while you're at it.'

He is buried in the garden cemetery at the home for retired Captain Hooks, in the amateur section, is my dear dad, and so that conversation will never take place. But then it is just one of the many that passed between us which in those days was as good as a glance.

The behaviour of terrorists in public

You hear things from other bus passengers, some things that ring true at the time of their telling but which later on reflection turn out to be not that reliable. Like the woman with a pile of canned cat food balanced on her lap. On every bend those cans threatened to spill. After I caught one she smiled and let me in on a little secret. She told me the cat on the cat food labels was actually her cat. Then she leant to whisper in my ear, 'But to tell the truth, Sammy can't stand the stuff. His tastes run to the more exotic.' I leant my ear to hear more. Well, just that morning there had been such an event, she said. She was horrified, she said. She was so horrified she screamed at the cat, 'bad cat!' and Sammy unlocked his jaws from the native bird and tenderly set about licking its wings and mashed head back into a state of repair. She got off at the next stop, and when she smiled up at the window all I could think of were the woman's wicked teeth and it was then that I had an idea of what those terrorists must look like and think about a moment after they have laid their explosives. I thought about the bird's mashed head and thought 'bullshit'. It was already a lost cause.

The story of the fish

Late night. A guy who used to work in advertising. Seemed to be on something. Talked on and on about point of difference, how vital it was to get your message across in today's world. That's when he opened his briefcase and showed me the fish project he works on part time. No one, he said, has ever painted fish on fish. It's a nice pun, don't you think? Is that what a pun is? Well, it sounds like it might be one. My idea, he said, was to paint these miniature scenes. Here, like this mullet. Imagine what he sees in the course of the day. A shark. A squid. A kelp bed. They're all there. The swordfish – I just put that in there for fun.

And after

When I got home from my shift, instead of going back to my flat I went down to the beach. I approached a diver and handed him the mullet with its painted scenes from the unseen depths of the ocean. I asked the man, 'Would you mind putting this back? Find a place where no one will ever find it.'

When globalisation came to our street

This is how it happened. First, our next-door neighbour returned from a VSA stint with a Rarotongan wife. She wore a flower in her dark braids. She introduced her smile, which was radiant and so different from anything we had experienced up to then. I knew the butcher's smile of satisfaction at a job well done. I knew my father's smile after he'd beaten me in a foot race. We always started at his say so, and he would decide when the race was over, usually – no, always, with him in front, his arms raised against the finish tape, which was invisible to my eye but plainly seen by him. Until the Rarotongan woman moved in next door most of the time we were grim as vicars, especially around the dinner table, along the sideline of the sports fields, addressing the golf ball at the tee. We never knew a smile that could be just a smile for the sheer hell of it. Well, that opened the floodgates. The Fischers moved into the next street and I made friends with Max, whose mother made me my first salami sandwich. I lifted the flap and peered in to find a couple of thin red discs studded with white beads of animal fat. I knew about fat on the rind of a chop or pork crackling, but salami did fat differently. Fat budded over this skinny flesh-coloured meat and I felt repelled, and yet I had already taken a bite and hadn't fallen down dead. Now I turned my attention to the bread. It was heavier, more textured than the fluffy white slices I smeared jam on. Our bread, I have to say, was more modern than the Fischers' bread. Our bread was pre-cut, a miracle of technology at the time, and more hygienic as it came plastic-wrapped. The Fischers' bread was large and round and sat on a breadboard. It looked alive, almost as though it might leap from the bench and pounce on the living. Fischers' bread was also hard work. You chewed and you chewed, you chewed and began to worry that you'd fall behind the white-bread eaters. We'd watched the moon landing at school that afternoon. Fischers' bread didn't have the same staginess and felt more real than those blurry images of men in bloated suits bouncing over what looked like a golf course at night.

Then, as if smiling and salami and chewing hunks of bread weren't enough, my sister returned from three years in Italy. Suddenly we were eating rice risotto – this was three months before the packaged stuff could be bought at Dennis's. Another sister returned from the world with an African spear for me and Venetian glasses for my mum, which were promptly locked in a glass cabinet and never seen again. We got TV that year. Dean Martin with his Las Vegas charm graced our sitting room. As soon as he began to croon 'Everybody loves somebody' my mum would smile – a smile I had not seen before, moulded by events from before my father stepped into her life to intercept whatever potential that life was unravelling. A smile that was a fossil of another person in another time.

Another day, the same old

The house creaks in the wind. My mum's head ever so slowly turns on its rusted hinges. Memory whips around the walls of the house and out the window. The storm is clearing. There are moments of clarity. A blue patch like yesterday and maybe tomorrow. We stand around, helpless. The house creaks and groans. Some old music is playing in her head; impossible to track. Something about a wind-up toy. She wakes at 3 a.m. and insists on drawing the blinds to 'look' outside at the 'day'. She goes back to sleep; wakes at midday and has breakfast. She watches last year's Wimbledon final, her jaw open and pocketing every excitement. A tie-breaker already resolved plays out (again and again) and the rest of us edge forward in our seats pretending we don't know the result.

Like death, I suppose, says somebody later.

We are between points.

We play with the remote.

Someone is mouthing off on television; guys with beards firing shots into the night over the city. Their elegantly dressed leader adjusts his tie in the studio.

Someone changes to the CNN weather update. For some reason that's where we stay: my mum with the same intense look of interest she showed in the Wimbledon final. So we learn it is 38 degrees in Honduras. Africa can expect fine weather. Rain in Singapore.

I get up and walk to the window. A sparrow on the window sill blinks up at me. I clearly interrupted something. Oh yes, the bisected worm.

Out on the road I can see a woman 'applying her face' in the wing mirror of her hospital services car.

'When I was last in Singapore it was always bloody raining,' grumbles a sibling.

The looping fly is far too quick for my mum's slow-moving head. We switch back to the tennis.

My sister opens a bottle of wine and that cheers everyone up.

I almost forgot to say

As I approached the window, with the weather forecast for Honduras blathering in the background, I had been thinking about Damien Hirst's shark in formaldehyde and the shoals of eyes that peer at it from every angle every day of the week. The last time I found myself surrounded by sharks was in French Polynesia. There were six of us guys and one woman on her honeymoon, and as she fluttered up the rope line to the dark shadow of the boat all of us guys were gazing up at her pale legs and her crotch. We were blowing our bubbles and gazing like a bunch of sad jack-offs at a peep show and all around us were sharks, perfectly still, like Damien Hirst's shark gazing at us.

Anyway, that was also the thought as the skimpy dress caught on the thigh of the tall, bare-foot woman boarding the bus this morning. It came and went. A flash of crotch, a vanishing shark.

And now, another day

My mum waits with her mouth wide open, like some ancient variety of sea bird. There are no words, no memories to feed her. She misheard. I was just clearing my throat. Now she must turn her head all the way back to TV and 'Dream Holidays'.

Parallel lives

In 1968, for the first time my dad was able to juggle his day-to-day exigencies (as he liked to call them; he was not an educated man and he had found that classy-sounding word in a newspaper and immediately taken to it), such as paying the bills, while wondering what he might buy my mum for their wedding anniversary, and at the same time take an interest in Elsie Tanner on the other side of the world; and from my bed, through the narrow gap left by the door standing ajar, I could see the rows of chimney pots and the closing credits that looked like lettering on tombstones and hear the heavy funereal music of a place that seemed weary of itself.

Clothesline

Then early one evening we walked out to the clothesline in a solemn line, my nana, mum and dad, my sister and me. Mum helped Dad peg old Nana up, then my sister; my nana seemed to know what was coming and so hung from the line as passive as a wet sock; my sister's big wide eyes held the wet grass beneath her dangling bare feet. It was that time of the year when dew quickly replaces the shadows laid down by the hills. And the sky doesn't know what to do with itself, whether to cling to the last of the sun along the hilltops or just give up and sink down to where its knees are already in dark. I took all this in as my dad pegged me up. I didn't see him peg himself up so I don't know how he went about it. Like my sister I was staring at the ground. I didn't feel so much airborne as held against my will. There was the brick house. There was the wooden fence. Now, as the clothesline began to spin, the brick house and the wooden fence appeared to merge. Now I could smell my sister's hair.

It was blowing back in my face. And as the clothesline spun faster I seemed to go where the smell of her hair led me. I seemed to wriggle into her being. I definitely felt something coltish pass through me. I felt like yelping and shrieking. Now Nana's solid form absorbed me. I seemed to slumber. I ached in every joint. And even though the clothesline was spinning so fast we could see the shadows of our feet flying in the windows of the sitting room I felt no joy. As I passed through my dad I felt a concern for the creaking in the clothesline. I was suddenly aware of a worn nut thread and a fraying line. I was hoping no one had noticed this. Above my sister's yelping I resigned myself to remaining silent. My mum's scented skirts flew back in my face. The clothesline spun at a dizzying speed now. Bits and pieces of my self flew off – my arms, my legs, my teeth, my eyebrows (that was a pleasing sensation, like painlessly lifting a sticking plaster and letting in cool air) – until I felt naked and hollow. The clothesline's rotations slowed and I found myself worrying about the lasagna I'd left in the oven. I worried about my nana and her depression. I was consumed by worry. Worries of every kind fluttered inside me. A thousand frets lumped up in me like hiccups. This was also part of the game. Whoever's skin we found ourselves in when the clothesline stopped we were stuck with until morning. Now I could see my mum was playing up. If she went on teasing her old mum in that way I'd have to threaten her with an early bedtime. 'Glenys! For God's sake leave the poor woman alone!' She looked up with an expression of horror – like a toddler on a school mat terrified of being singled out from the litter. I wagged a stern finger. 'Be nice,' I said, and as my dad came and stood by me in a gesture of parental solidarity my mum looked on helplessly.

A song my mum wrote on the back of an aerogramme

Under the skies of the Maniototo
I felt myself return/ I felt myself return/
Under the skies of the Maniototo/ I wondered where that self had been/where that self had been.

The end of small time

My father signalled his death with a crashing. My brother's girlfriend and my mum rushed upstairs. I learnt all this later. My father lay in the door of the bathroom. For some reason my mum thought to undo her husband's shoelaces. The girlfriend applied mouth-to-mouth, which is rather what my father would have wished. My mother's own death was signalled a year in advance. There was no crash. Waiting and patience would be required. We took it in turns to sleep over in the ward. Slumped in the hard chair, I felt like I was flying economy over the Pacific. My head rolled around on my shoulders; every so often I snapped to, and I'd find my mother with her hand raised, pointing. She'd done the same thing the morning we broke out of hospital and I pushed her in her wheelchair along High Street; coming in the other direction against the flow were me and her, she was holding my eight-year-old hand, and we were walking to the library to get out a book. At Janus's I bought a latte for me and a cup of tea for her – not that she drank any of it, but at least the cup of tea signalled that she was back among the living, even if it was just a small cameo appearance. It was the hottest day of the year, cloudless, blue. 'Look,' she was pointing, but I could not see what she thought she could see. Through the night I got up and walked over to her bed. At least a dozen times I re-introduced myself. Otherwise, I sat in economy class listening to the noises of trolleys, squeaky wheels, the ringing sounds, buzzers, and just before dawn the noise of a vacuum cleaner crept along the polished floors. I pulled back the curtains. My mum turned her head to the new day. 'Oh God,' she said. She sounded so miserable.

She has other voices, though. A whole backlist that have never left her.

There is the timid voice of the girl sitting up to the piano, her knuckles raw from punishment for hitting the wrong key; the petulant voice of the young mum resentful of the husband's coming and going from the fortress of crying babies and endless nappies; the more tyrannical voice which happily ceded to a kinder, more effortful voice in late middle age, the kindlier voice that likes to kid along; the whimpering that is the right of the very young and the very old. None of these voices ever departed, it seems. And worse of all, a voice I have heard twice now, a whole new voice of terror, of a how-could-you–do-this-to-me kind, as we wrestle her mouth open to take the morphine.

The end of big time

In this morning's papers, more provocative talk from the Iranians. The Israelis do not rule out a pre-emptive strike against Iran. From the Chinese a warning to the Americans not to meddle in Taiwanese affairs. A Chinese general considers a pre-emptive strike against the US within his country's rights. And a teaser on the cover of a prominent American magazine promotes a story about how to defeat the Chinese in a future war. In the same news bulletin, Genesis has left Cape Canaveral. After seven hours it will have passed the moon. By the year 2015 it is expected to reach Jupiter. The possibilities just keep on expanding. We live. We die. We live. We watch out for the hairy sun.

Gridlock

The traffic is backed up as far as we can see. Shiny top after shiny top. Now and then an opportunist darts into a gap. Youth. We say it with a shake of the head, a roll of the eyes. The predicted wind has failed to arrive. The hills. The riffs of quiet cloud. We are all waiting. It is hot, so everyone has their windows down. Everyone is tuned to a different radio station. In the midst of the rock station hysteria comes a violin, delicate, insistent. There is gunfire and bomb explosions from a news bulletin. Some canned laughter. *Hey, I'm just an ordinary guy! No, really.* There are advertising jingles. It is hard to tell them apart from the pop jingles. Which, of course, is what advertisers like, as well as, presumably, the US State Department, which has just admitted to a policy of disseminating false information, spraying it out there like weed killer to burn off trails to the truth or maybe flush out a mad man, an assassin, a hijacker. The noises of the world are no longer reliable.

The driver of a van in front gets out with a resigned look. Long hair, whiskered growth, a rock star's mo. Faded blue overalls. I watch him dig around in the back of his van, then he turns and, finding my face in the queue, holds up a beer can and points to it. What the hell. I get out of the car. His name is Frank. He's a furniture polisher. The beer has been in the back of the van a bit long, but hey, this is better than sitting marooned in a line of cars. While we stand there drinking the driver behind me gets out of his car and walks sweatily towards us. He wants to know if he can borrow someone's mobile. His battery is out. He's late for a meeting and needs to send word. I give him mine. He turns away but we can hear him clearly. He very definitely sounds middle-management. The way he talks up the problem. For the benefit of the other motorists there's a bit of posturing. His gestures are unequivocal. He cuts the air with his hand. He signs off with a Latin thrust of the arm. He passes me back the mobile, and without any hesitation accepts a can of beer from Frank. It's been a helluva day, he says. Crazy. Absolutely crazy. His name is Graham – office systems. He says what that is but neither Frank nor I listen. We sip our beer and watch the traffic shuffle up one place in the outside lane.

Soon a banged-up Subaru nudges into the vacated space. We raise our beer cans above our leery faces but our buffoonery hardly registers with the man. His hands are stuck to the steering wheel as if he might be going somewhere. I don't know why he can't just abandon ship and slip out for one of Frank's beers. We are all giving him jeering looks when his body slumps forward. Frank looks at me, and thoughtfully strokes his moustache. Graham calls out to him, 'Hey fella. Oi.' He taps on the passenger side window. Nothing. He opens the door and leans in. He reports back, 'We have a dead man here.'

The traffic in our lane shuffles forward, and as all the traffic banked up behind has just seen that fresh land they start honking their horns.

So what we do is this. We push the dead man's car to the side of the road in front of Frank's van. We are wondering what we should do next; there are obvious options and responsibilities, like phone the police or an ambulance, but is there any point just yet while the traffic is gridlocked? Our decision is delayed for the umpteenth time when it starts to rain. But for the dead man we'd thank Frank for the beers and head back to our cars. But we can't leave the dead man there on his own. So we pile into the banged-up Subaru. Frank sits in the front with the dead man. I sit in the back with Graham, all stomach and short knees. Late 40s, I'd say. We talk about what to do. There's not a lot we can do. We are stuck. We wonder if we should try and find out the man's name, but Graham makes some dark mutter about tampering with dead bodies. Frank turns around to see what I think of that. Actually, I don't have a problem. We're not going to rob the man. We just need to know who he is. Then what happens is this. The man's mobile phone rings. It's on his person; possibly in the jacket pocket. Frank lets it ring. He waits for the sound of a voice-mail message, then he plunges his hand into the dead man's pocket. He's over the consulting thing. He retrieves the message and holds the phone up for me and Graham to listen to – a woman who is obviously pissed off. 'I waited for you by the gate. I can't wait any longer. Katie's show begins in 10 minutes. I can see her and the other kids looking around for me. In case you've forgotten we're at the park entrance.' In a more facetious voice she names the park, reminds the dead man (we have a name at last), Joe, what city this is, the country and hemisphere, and what time he is expected. According to the time on the dash he is 10 minutes late.

Well, a few minutes later the traffic begins to move. We move the dead man, Joe, into the passenger side of the back seat. Along the way there are some funny looks from passing motorists. Joe doesn't look too good. I've leant him against the door in the back, his head against the window. He's not a good colour. He looks mildly angered by something, perhaps slow service in a restaurant, something of that order. Graham and I park our cars on the shoulder behind Frank's van, then we cram back into Joe's car and continue on into town.

Frank drives, mindful of the speed limit. We find the park. There's the gated entrance the woman spoke of; we drive to the circular green at the end. There's some sort of nativity play. Kids with cardboard swords, in costume, a gold crown here and there. A mob of parents in a proud semicircle. The women talking to one another behind their hands. One or two of the men are nearly falling over with boredom.

I'm last out of the car. I make sure I bang the door shut, and a woman – in a light, summery cotton dress, a bob of dark hair – turns and looks back in our direction, at first without much interest, perhaps it was just to see where the noise came from. But now we see the mystery catch in her face. She knows that car. She doesn't know us, of course. The questions line up to be answered. Who are these strange men driving up to her daughter's nativity play in Joe's car? Who are we? Why are we here? And where is her husband? Where is Joe? This is the moment Frank steps away from the back window.

Gridlock 2

I can't go out. Bombs are going off all over Jerusalem. On CNN the *Tampa* is stuck in a still ocean. The cameras hover for a look at the refugees huddled on deck and we are reminded of another boat with human cargo that no one wanted. I am stuck in the barber's chair. There're just the three of us – myself and my rather disappointing reflection, and the young guy with the braided hair and almond eyes whose name is Aziz – and when I ask (as we all do) how he got here he says, 'Aboard the *Tampa*.' I think back to that tiny hotel room in Jerusalem. A Saturday afternoon. Flies pouring in and out of the open window. A warm breeze steady on my face. 'Sweet,' he says later when I pay him, *sweet*. He is one of ours now – stamped and delivered. We like to band everything we catch: sea birds, fish. Each morning he wakes and rubs away the dream of the sun-baked decks of the *Tampa*. On a departing note he asks me if I am going to the rugby Sevens. I can hear the sirens all over the walled city. It is some mad fan in the crowd, and the men in black have just scored. The curfew won't be lifted until dawn.

Gridlock 3

We drag our net in to find it contains an elephant fish, several starfish, a sand shark, six tarakihi, a snapper, a stingray and a flounder. In our street live people from Ethiopia whose night dreams are filled with sand dunes and whose bellies ache with certain memories. There are people from Samoa; the oily black smoke is from the pig they are singeing in their backyard. There are Brits who follow Chelsea, others who wear Man United scarves, and some for whom the long-running *Coronation Street* is an ancient song going back to the beginning of their creation. There are families from Kiribati who walk slowly and carefully along shopping aisles as though picking their way around bits of sharp coral. There are Laotians, Thai. They fill their carts with dried seaweed. There are people for whom the supermarket is still a mystery. They don't know whether they are in the pet food section or baby food section. There are labels of leaping fish and smiling cats. When I look at what our net cast up the wonder is that such a variety came to be swimming in the same place at that particular moment when the slack tide turned.

Rings of Saturn

She raises her hand to ask permission to go to the toilet.
She raises her hand to ask permission to use the facilities on the premises.
She asks permission to enter the gardens after hours.
She asks permission to use the phone, her car has a flat back on the highway.
She asks permission to borrow the umbrella.
She asks permission to use the car.

We keep being told that it is okay to give her our permission to die.

Rings of Saturn 2

In the video game of the same name you have to blast your way through the ever-encircling foe. To encourage players, every so often – usually after a bloody battle in some clearing or on a rocky slope – a barefoot woman in diaphanous robes appears. A light breeze is always fanning her face. And when she turns it is your trigger eye she catches, always. We re-holster our weapons, step over the corpses, and carry on. This woman who has no name keeps us on the straight and narrow. Some of us have our own private name for her.

Meanwhile, the cartoon wars continue in Copenhagen and Tehran. The cranes are gathering in Siberia. Before our video games we shudder with excitement: the foe lie strewn over the battlefield. At midnight the cranes fly over Tehran. They have no opinion, they bear no grudge. The cranes return to Siberia without news, or memory. We sit on the school bus. The cartoonists lie in their beds. All of us wondering how best to strike our target.

Rings of Saturn 3

On a clear night it is a pretty picture. An ancient presence circled by caregivers.

Rings of Saturn 4

She is sitting on a hay bale. Her round schoolgirl knees are smooth and tanned. Ricky takes aim with the dart and misses. He can't get past four. I am already on 15. We are playing round the board. First to hit the triple 20 gets to kiss Paula. Deep in the groundsman's bunker Paula sits demurely on the hay bale, a schoolgirl Helen of Troy. Ricky is a sweating wreck of misery. He should never have challenged me because I always play to win. He loves Paula, and I am just some Johnny-come-lately tosser who has climbed in on the act. Still, a game's a game. I can see Paula is preparing herself for the inevitable. She smiles up at me, and that's when I see something. I see something and mysteriously – as far as the other two are concerned – I can no longer see straight. My darts bounce off the wire. I am stuck on 17. Ricky comes racing through the teens, overtakes me, bounces on to 20 and bulls eye! He's done it, and believe me, nobody is more pleased for him than I am.

Now she sits across the café table, some grey roots showing in her hair. Still attractive, more attractive in a way. Her eyes sit deeper. They are still, and not quite ready to reward wit or humour as quickly as they used to. She has come to hear me read. Over a drink she has nothing to say about the reading – the trials and tribulations of a hotel clerk on the night shift – but wants to know if I threw that game of darts against Ricky, whom she went on to join in an unhappy marriage, no kids, before discovering some late-to-surface truths about Rick, who she says these days runs a popular gay bar in Auckland. She wants to know what caused me to lose my way all those years ago, what did I see? And all these years later I still cannot tell her that what I saw was a tiny piece of lettuce leaf stuck against one of her front teeth.

Rings of Saturn 5

We say to ourselves, 'What a fantastic night sky.' We always know where to look. We look to the bits in between. We lie on the wet grass looking up. We lie beneath the palm trees brushing their palms away with our eyes. We lie in the backs of utes outside noisy pubs. We lie there trying to find which part of the world we've fallen out of and which bit we would most likely fit into now.

Pop charts

The palms flutter in the breeze. The colour of rust on everything, the flat rooftops of Nuku'a'lofa and the hulks sinking in the shallows inside the reef. The king eases his considerable weight forward and plays the first bars of Bach's *Goldberg Variations*. That is all he knows. He will hum the rest. He will hum Bach through his morning bath and later while sitting in the convoy on the way to the gym. Kings, princes, chess geniuses, life prisoners, architects of genocide, generation after generation from the People's Republic of China, from Russia and the Central Asian republics, up and down and across the two Americas, continental Europe, down through Asia and across the islands sprinkled over the Pacific – the audience for Bach just keeps growing.

Pop charts 2

The song goes back in time. No one knows how far back. But for the present moment the tui sings from a branch for an audience of one, a cat which has climbed up onto the roof of the parked car to get closer to its idol.

Pop charts 3

The stars are the true artists of our time. When they come out at night they do not seek applause. And only rarely (in the case of mitigating circumstances) do they keep their audience waiting. Sometimes, though, I think we fans read too much into their presentation: a chariot here, a lion there. Sometimes a collection of stars is just that and nothing more. I was still quite young and ill-prepared when I first learnt of their sad inevitability. It was Guy Fawkes night. We waved our sparklers about in the dark until they went out. In the morning we found their cold grey soot in the wet grass.

Travels with George

Directly outside the bach the moon casts a silvery reflection on all things that float, and as kids, on Uncle George's launch, we used to putter from silver canyon to silver peak, across her deserts, and whenever we were late arriving home and Aunt growled, where on earth had George been with us kids, up at this late hour, he'd say, 'To the moon and back.'

2

But sometimes we'd just stay in the backyard with an eye glued to the telescope and the living room windows thrown open to the soundtrack from Stanley Kubrick's *2001: A Space Odyssey*.

3

One night I borrowed George's dinghy and rowed my first love across the plains of Venus. But, alas, she did not know how to play the game.

4

This time we have come aground on a sand bar. With a certain amount of glee George announces, 'We're shipwrecked.' The wind blows up. The night comes, and so does that old nightmare of something creepy hiding under the bed. Tiny microscopic life-forms are nibbling at our sea-rotted feet. You lift your foot and stare, but you can't see anything. Stars – there are a multitude. So we play 'I spy'. We join the dots to create sketchy outlines of loved ones, and in George's case of favourite buildings in foreign cities he's visited. There's the Empire State Building with its steeple. We are all struck with vertigo. We look down at our bare feet. We stand on the driest part of the sand bar. The tide slips and slithers up our shins, like it is measuring us. The bogeyman is looking for us. 'Montana,' says Uncle George, out of the blue. 'Now some believe that to be a land-locked sort of place. But once an ocean filled Montana. You can still find old snapshots of ancient fish stamped onto rocks. There were sharks in Kansas. Crabs in Utah. Bony fish in Oklahoma. Idaho even had a beach front back then. It shaped onto the Skull Creek Seaway.' It helps to know that stuff. It helps to know that despite the incoming tide our feet will still touch land. And vice versa. Because at full tide we will slip back to deeper water and make it back in time for hot drinks and some fishy tale from Uncle George for my mum. Meanwhile, we are set the task of finding the Kansas shark up in the night sky. I can't see it. I am stuck with the Utah crab. Everywhere I look I see crabs.

5

I sit in the outer waiting room of the palace in Nuku'a'lofa. I have come to interview the king. The protocol is clear. I am to ask only inoffensive questions. With the help of the king's secretary I have prepared a list of potential topics – boating, royalty, renewable energy, space exploration, and Bach, His Majesty's favourite composer. I know he is prepared because when I arrived I could hear the first bars of the *Goldberg Variations* trickling out the open windows of the upper floor of the old wooden palace on the waterfront. Now, the secretary pops back into the waiting room. He looks worried, more worried than the other day when we met for the first time in his waterfront office. The folds above his eyes have deepened, and his forehead is moist with sweat. He has thought of a thousand things that might go wrong and has experienced them already like they just happened, one calamity after another. He explains that HM will be along in a few minutes. In the meantime, would I like to look at his moon rock collection. I might as well, so he shows me into the exhibits room. The rocks look like any other rocks heaved up by a cold river, but of course these are from the moon and so you find yourself looking harder to make sure you haven't missed anything. You want everything of this remarkable contact to be scoured into memory. You are preparing for future conversations in bars across half a dozen continents where you will reach for this moment: did I ever tell you about the king of Tonga's moon rock collection? There is something else reflected on their silvery surface. You stare, half wishing, half hoping for an insight into the Divine. The creaks on the stairs are HM. And at the creaking the secretary raises his eyes from the moon rocks to the ceiling. The creaks on the stairs are the most important thing he has heard all day. That creaking is more deafening than the dawn chorus of dogs and roosters. I think how there have been other creaks on the stairs, equally profound, even fearful, made by this king, that dictator, and I pause to think about that extraordinary moment when Neil Armstrong stepped down the ladder onto the surface of the moon. As he did so, as I recall, there was no creaking sound. He dropped onto what my father said looked like the sand trap on the 17th over at the Hutt golf course. But now the king is coming and I must pay attention. Some older men in black shirts and matted skirts kneel on the floor to prostrate themselves. No subject is to be higher than the king. I look around for the secretary. He has the key to the glass cabinet holding the moon rock collection. He is filled with new urgency. He takes the rocks out, one after another, piling them into his arms. He looks around for help so I take some out as well and do as he does; I spread them over the floor so HM may walk across them. So few of us have lunar-walked; so few of us know how easy it is.

Time delay

1

He checks his watch. He looks up. She is not turning up after all.

2

The crowd cheer the dwarf. They cheer and cheer, until it dawns – the hippo has closed its jaws. The hippo has closed its jaws and there will be no fantastic escape as advertised.

3

It's the father we remember, his face raised to the blue skies, his proud teeth – yes, even they are proud, he is so proud, proud, proud. He is so proud that he goes on being proud even after seeing what the rest of us have just seen on our television sets: the ball of flame, tapering and flaring; what was just a second earlier a space rocket pointing and moving in the direction of the stars with his daughter, the astronaut, on board.

4

He checks his watch. Then, as he looks up, there she is. She says, 'Look at your face. What were you thinking?'

What the second-hand mirror knows

1

The grateful face of the school principal loosening his tie after the last parent has left his office.

2

Eichmann brushing his teeth on his first morning in Jerusalem.

3

One last look by Bob Dylan before he goes out to face the audience of concert number 5092.

4

The tour operator five minutes before the group is due to meet downstairs in the lobby at 07.25 on Day 12.

5

Uncle George's 15-year-old daughter Ness waiting for the results of her pregnancy test.

6

Pope Pius IV finds a pimple.

What the bathroom mirror saw at 29 Pacific Drive

The wife sitting in the bath frowning up at the moley back of her husband.

The wife and husband sitting in the bath facing one another, their knees bridging the foam, but their faces too far apart to see one another.

The daughter of the wife and husband sitting in the bath with her boyfriend smoking ganga.

The African woman in her green uniform bending to bathe her elderly father.

The real estate agent and the African woman's husband in the bath.

The African woman and her husband, two black nights.

The candle-lit faces of the African husband and the real estate woman.

The small mulatto boy shoving his boat across the foamy ocean, admired by the real estate woman.

The bereft face of the real estate woman, the red eyes of the African man.

The real estate woman's body slipping under the waves, the candles burnt down to waxy stubs, the empty pill box, a man in overalls stamped with the word CISNEROF.

That old story

The old man is looking out the window, enraged by something, a section of newspaper in his hand. Even though my view of him is cut off at his waist I know he has his slippers on. And I also know what he's thinking: what the hell are we doing sand-hoppering around the backyard? Why dad, sir, we are trying to jump back into our shadows. It is not easy. Not as easy as that face in the window may think. We go back to the problem. My friend Max Fischer says he has an idea. We need to outwit our shadows. We need to think of something else; better still, just act like we're doing something else. The idea is to bore our shadows, then pounce when they're not ready for us. So Max climbs a tree. I walk along the fence line. For a while I stare at the bit chewed off by the handicapped kid down the street. I look for tooth marks. I watch a huhu bug stick its head up into sunlight. It must be an insomniac. Do huhu bugs sleep? Max will probably know. I wonder what Max is doing. I squint up into the tree where Max sits dangling his legs – now he works them like a pair of scissors, and when I drop my eyes to the grass I see what he's doing and my heart sinks. He's trying to kick free. We simply cannot shake our shadows; we can't even stop thinking about them. The only thing left to do is to punish them, and this we achieve – at last! Bastardos! – by dragging our sly selves to the garage side of the backyard, thereby sinking our trailing shadows in the deep afternoon gloom cast by the garage. Max picks up a stone. He's thinking about stoning his to death. But something makes him look up at the living room window, and seeing my old man standing there shaking his head changes everything.

2

She is very attractive to follow up the stairs; narrow in the waist, her dress flares out like one of those 1950s newspaper advertisement models showing off a new refrigerator. We count the numbers down the hall. Most people flop into a room. This woman marches through the door like there are still things to accomplish at this late hour. She makes her way to the window facing the square and there she pulls the curtains apart. People who know this hotel know to ask for a room on the other, quieter side of the wing facing the square. But she was adamant about room 202 so she's got what she wanted, that, and a few late night drunks tottering across the square, and a noisy rubbish truck and a last turn by the street sweeper. I need to get back to the front desk. But as I start to leave she holds out a hand, 'No, no. Stay. Please.' She looks around the room for the first time. 'Stay for a drink. Just a nightcap.' And when I hesitate she says, with a small laugh, 'That's all I'm proposing.' In a jiffy we have our brandies and the woman quickly takes possession of the room. She points to the armchair where she wants me. She takes the two-seater for herself. I quite like the way she perches with her knees touching. I am reminded of an old lost opportunity. She has a nice smile. I think, if I knew her better, I would tell her to smile more. As I surface from that thought I catch her spying on me, half fascinated in an eye-to-the-keyhole sort of way, and half amused by the thought of exactly that. She says, 'You will never guess why I am here. Not in a million years.' She jingles the ice in her glass. 'Something happened in this room a long time ago. Something which I am still struggling to come to terms with. I could have said "understand" but I think struggling to come to terms with gets to the truth of the matter with something more than just understanding, if you see what I mean.' When she smiles up at the ceiling I see what lovely teeth she has, all neatly stacked, whiter than white. She puts her drink down on the table. I do the same and follow her over to the window. Glancing down at the carpet she says, 'Do you know, where I am standing my mother once stood, right here, before this very same window, and down there a man walking across the square stopped, looked up, and waved. And then, the rest of it, well, I don't know how it would have happened. Even when I stand here I can't fill in the gaps.' For the moment the two of us stare down at the empty square. I don't know what to say. I don't know what she is talking about. In fact, I don't know why she wants me to stay. Then she says, 'Now if you were to go and stand down there in the square I might know what my mum felt, what she was thinking. Or is that too weird?' It is weird, very weird. But she touches my arm briefly and I like that.

There is a sharp autumnal chill in the air. I noticed it at the start of the shift, too. I thought I detected a whiff of playing field mud, a sweeter smell of cut grass. Some old memory of that hour and lining up at the bus stop with my muddy knees and a lightness in my over-worked legs. As I walk from the hotel I remember another time when I am walking along the stopbank. The river itself buried beneath a long, skinny wreath of white fog. I have missed the last train, and so long as I keep walking I won't get too cold. Across the valley the squeal of tires, the sound of broken glass. The birds wide awake and still in the trees. The dogs standing at attention in dim backyards from one hill to the other. The girl I stayed with who caused me to miss the train, I hear she married an alcoholic. There is a photo her older sister took of her doing a duck dive in the Aegean. Her bare bottom tucked up and a dark cleft between her thighs. She loved to show me that photo.

The drunks have left the square. A few streets away I can hear the sweeper. I turn and look up. The hotel looks wrapped up and unavailable. In all the windows the curtains are drawn, except one. There's the woman in 202, her hand raised. I copy her and raise my hand. When she presses her hand against the glass and turns her hand onto its side, spreading her fingers, I do the same. Now she moves her hand back towards her chest. She looks like the lighthouse keeper's daughter – I am reminded of one of those long cutaway shots of a woman left at the window following someone's departure. But this is different. No one has departed. So maybe I have that scene wrong. Maybe it is more like one of those English period numbers. D.H. Lawrence, that sort of thing. The kept woman with the powerful urges at the window, the brute gardener working his pitchfork in the hay. Or maybe it is just two divided things seeking unity.

Later, much later, as I am pulling on my trousers, already filled with guilt at leaving the front desk unattended for so long, she says, 'That man my mother waved to, he ended up my father.' She is lying back tangled up in the white sheets, her head turned away to the past event. 'And guess what? He was on his way to table tennis when he stopped to look up at that window. Table tennis, for goodness' sake! How weird is that!'

3

The American trucks pass our house on their way to the river to pick up shingle. I am not yet born. I am still just a twinkle in my parents' eyes. A future project, shall we say. My older brother is around. He's just been popped inside the back door with strict instructions not to leave the house. 'I won't be long,' our mother says. He hears the key turn and races to the front of the house and looks out the window to see her climb up into the truck cab and ride down to the river with the Americans. My brother says there are some things you don't forget at four, so who knows. This is on the phone, 60 years after 'the event'. He says there were tens of thousands of Americans stationed here. Our guys were away at the war, or, in our dad's case, down at the wharves patching up blown-apart American ships all hours of the day and night. My brother says that at the end of the war the Americans had to rent ships – it's true! They had to rent ships! – to take all our women back to the States. Now, as they get older, lots of them are coming back, some to die. What prompted this dialogue was a photo in this morning's paper. Some canny old ex-pat bird with a pile of hair and a cheap necklace now living in Nevada, towing behind her a skinny, retired-looking guy with a stomach. 'That could have been Mum,' he says.

The drawn map

There is a place where all the currents meet, and containers fallen off ships spin slowly through a soup of plastic and packing wood. There is a place in our heads where all the bits of life circulate. There is a tip face where, if I like, I can find my old sheep, Bert, in all his incarnations – a lamb prancing around lifting his stubby tail. He got dumber with age. There is the jersey my mum knitted from his fleece. There are the slippers my dad flung at Bert from the sitting room window, furious at losing sleep and right smack in the middle of the season as Hook. Bert is a universe of meat and wool, of consumption and baa . . . ch. He is an epistle of devotion. He followed me to the shearing shed, and on to the butcher's. I wore Bert, and there I am eating him, seated at the table, surprised to hear of Bert's legs spoken of as French cutlets. Besides my jersey, Bert ended up a yarn that tangles and disentangles according to which stage of his living I reach for.

At the tip face I can find the hotel guest standing in her mother's shadows; there is the ping-pong player – a faceless man I thought, until I looked and saw myself in the mirror; there are accidental encounters, a man called Joe about whom so little is known, but I will never forget his angry diner's look; there are old moments with white creases; there is the costume of life, some of it flung about in any old fashion. There are things that contain more than meets the eye. There is a stain of fur on the road, once a cat with a blackbird in its mouth and a thousand and one memories of having crossed the same stretch of road in rain and sunshine. There are all the yesterdays and todays rebounding away. There is the chewed pencil end used by Einstein. There is the toothpick used by the king of Sweden. There is the slingshot – of incalculable worth. There is the distinguished visiting physicist leaping in the air and clapping his hands and at that same moment squealing in an excitable Scots accent: 'Space-time!' There are scraps and bones, still, bereft as if awaiting a dream to absorb them. There are and there is, and a lot piled on to be picked and sorted; and you can find anything, absolutely anything at all, as Demeritus discovered, by simply joining the dots in a whole new way never seen before.

EVERYTHING WE KNOW

JO RANDERSON

*That which arrives in pieces will be
bound together into a whole.*

NABULAN PROVERB

The beginning – the myth[1]

There was a time before the time which was the beginning and there was a time before the time which was before non-being. Before heaven and earth took shape, there was only undifferentiated formlessness; the heavens had not been named, the waters mingled as a single body. None of the gods had yet been brought forth and the primordial abyss was everywhere, stretching endlessly in all directions.

And then, suddenly, there was the beginning, as if from nowhere. A single moment! One little push of feet and then the heavens, the earth – something was moving over the face of the waters: Form! Colour! Differentiation. *Something was happening.* And someone said, 'Let there be light', and there was light. And the light was called Day and the darkness Night. And there was evening and there was morning, one day.

And this was good! At least someone felt that it was good, and so there was another day, a second one, and also a second night to maintain the balance. And then other binary concepts started to create themselves: there was winter and there was summer, there was yin and there was yang. There was energy and there was matter, there was science and there was art, and there was truth and there was myth. And the Line-That-Separates-All-Things was made; we ate of the apple and learnt the power of discernment. Language was born! The Great One became the Many Different. War followed soon after.

The beginning – the truth

A few decades ago, the truth was that there never *was* a beginning. At that time it was true that the universe had always existed and would always exist in the same steady state that it exists now. Nothing

1 Taken from Egyptian, Christian, Taoist and Maori creation stories

was changing, except that everything was getting constantly bigger due to the continuous production of new matter, but it wasn't really *changing*, which was comforting for those of us who like things to stay as they are. For a while, this was the truth, a slightly duller truth, but nevertheless a truth worthy of respect like any other.

Then came a new truth: there *was* actually a beginning, a rather large one. There was a cosmic egg which then exploded – there was a rather large bang. Then everything started moving outwards, steadily increasing in size, which it will continue to do until it eventually reaches a point – and no one is sure exactly where or when that point is but the point will know itself when it gets there (it will reveal itself in a blaze of glory, a loud fanfare, some sort of bang or whimper) – anyway, it will be clear that it is the appropriate time to reverse direction and everything will turn back in on itself, and then begin moving inwards while continually getting smaller until it reaches an infinitessimal state, a tightly packed nugget of the densest possible matter (another cosmic egg!) which then may explode again and start moving outwards, until it turns inwards again, continuing the cycle ad infinitum.

If the truth sounds as unlikely as the myth to you, let me briefly explain how we know this. We *know* this because of the dark matter, the antimatter. Everything has its opposite, its shadow-force, its negative image, and there is obviously matter, because we can touch it, we see it and we know it, so therefore there must also be antimatter, which we can't touch, can't see, don't know, but it is there, *we're pretty sure*, well, we're certainly pretty clearly definite that this is a very strong possibility. At the moment that's what we think.

Here's another definition that might help. This is how Stephen Hawking (a very intelligent man with a unique mode of communication) describes it: 'At the critical moment of creation, a mysterious group of particles broke free from the pack. They began to cluster together to become the dark matter. These clusters of dark matter had a gravitational pull on the ordinary matter. As the lumps of dark matter grew bigger and bigger in the expanding universe they pulled in more ordinary matter which condensed to become stars.'[2]

2   From 'Stephen Hawking's Universe'

What caused those mysterious particles to break free from the pack? Why did they begin exhibiting this abnormal behaviour? Why didn't they just follow normal procedure like the other matter was doing? We can't answer this right now, but for whatever reason, that's what the mysterious particles did, and we are lucky that they did because it set a whole chain in motion.

I'll explain it the way it's been told to me. Stars formed, and other things happened too, things *changed*, which can often be upsetting and painful – I'll skim through quickly here just touching on the highlights: there was some primordial soup, an amoeba, one giant land mass; enormous amounts of ice came and went, as did some very large animals. Then we were pulled out of Adam's rib (Adam was one of the apes) – they took one of his ribs and fashioned it into the shape of a lady. Some fire came down from a mountain – we got thumbs! This sped things up significantly, so that some of us, some of the male ones, went fishing in a giant waka[3] and this land that we're standing on now was at that moment a very large fish under the water and the fish got caught and pulled up and now it is the land. An island. It's the North Island or Te Ika a Maui, the fish of Maui,[4] and that's what we are living on now, so I hope that explanation has cleared up any uncertainty as to how it all began.

A question

When I first started researching this project my question was, why do many natural phenomena look the same? For example, why are there the same patterns in the topography of land, ripples on a pond and clouds in the sky as there are in the folds of human skin when you squeeze it together? Why do a flock of sheep cluster in similar groupings to a copse of pine trees? Paul Callaghan, physicist and communicator extraordinaire (one of my appointed collaborators for this project) said, 'You're talking about self-organised criticality.' I was grateful there was a name for it. But before I discuss SOC and the ethical implications of this concept, there are a few other scientific principles I need to cover briefly. This lecture is like a flock

3   A Maori sea-faring vessel, like a canoe.
4   In Maori mythology this is how Aotearoa New Zealand was formed.

of pigeons and my goal, rather than caging them, is to liberate them and observe the patterns as they flutter out of sight.

Central to our study of science is the concept of relationship. Not only *what* is an object, but *how does it relate* to the others. How do the planets move around each other? Who moves around what? What is the relationship between these giant bodies? When we study the world around us, we must look not only at individuals but also at the environments in which they are placed. Copernicus's great revolution was the discovery not that the earth was square, nor made of cheese, but simply that it orbited around the sun, rather than vice versa. A relationship can be revolutionary. It can even lead to death threats, or just fewer party invitations.

And many of us may say, 'Relationships, who needs them? Why would you bother developing any sort of relationship with someone who is just going to lie to you, cheat on you, take you for granted and take whatever they want from you, never really caring about you, professing to love and adore you but failing, constantly *failing* to anticipate where your real needs are, well, forget it, wouldn't it be better just to exist on your own as one little individual looking after yourself and wouldn't our society be a lot stronger if we all lived like this, because a rock, as we all know, feels no pain and an island never cries?' Some of us may say this.

But this is not how nature works. No man is an island, nor are very many women, and I would like to introduce several concepts by way of illustration. We know from studying all organic systems that with time, the development of relationships is inevitable. Entanglements occur. Structures move from simple to complex, from order to chaos, from uniform grey similarity to infinite colourful variation. This is the nature of life, and it is not only a spiritual and poetic concept, it also has a scientific name: **entropy**.

Entropy is the measure of the disorder of a system. All systems, with time, will increase in entropy; that is, all natural structures move from order to disorder. And to prevent this happening we have to put energy into the system, we have to do 'work' to keep a system ordered. The commonly-cited example is a teenager's bedroom, which generally requires a mammoth clean-up every few weeks, only to degrade into chaos as soon as parental backs are turned. (Perhaps the creative psyche works similarly.)

So this is the concept of entropy. Any structure left to its own devices will naturally form a disorganised state. The tendency is towards messiness; complexity and messiness. Anarchy rules! This is the pattern of the organic community, the entangled, disordered, interconnected state. This is the observed reality of all natural systems.

However, this large, complex interconnected state is made up of small, finite individual units. Let's consider quantum physics for a moment, but not too closely, let's just look at it out of the corners of our eyes because, like the sun, if we look at it directly it can be disorienting, blinding even. **Quantisation** is the idea that the natural world is granular rather than smoothly continuous. Like a bowl of sugar rather than a jelly. Like a pixelated photo with an overall smooth effect, whereas under close inspection each individual dot is visible. So in this way of thinking, we are aware of the importance of the individual, the single unit, the solitary grain of wheat.

Furthermore, let us briefly visit **the uncertainty principle**, which concerns the reality that an electron can exist both as a particle and as a wave, i.e. both as matter and as energy, and that at any one time, *because we are watching it* the electron could react in any number of ways and we can never predict exactly how it will react, although we can make some guesses. We can predict a range of possibilities but we must factor in some *uncertainty*: basically, how far we can stray from the path but still remain within cooee. From quantum physics we observe this possibly surprising form of natural behaviour – individualised units which exhibit unpredictable behaviour but within a certain range, clearly affected by the surrounding observers and environment. (If this confuses you, don't think about it too much. It might be easier to understand just by looking at a bowl of sugar.)

So, combining all this together, we have a natural pattern like so: *complex interconnected systems consisting of generally (but not precisely) predictable individuals where each individual's action affects the behaviour of the others.*

This is a good place to introduce self-organised criticality. It works as follows:

Say you are measuring earthquakes. You will find in your data – and this is not rocket science – that there are many small tremors, a moderate number of medium-sized tremors, and a small number of very large tremors. And you get a graph result that looks like this: many small, moderate number medium-sized, few large.

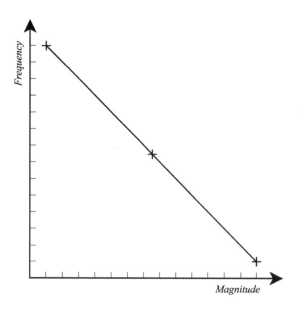

If you're better with words than graphs, here's another version.

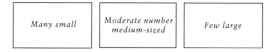

And here's a picture if you're an abstract visual learner.

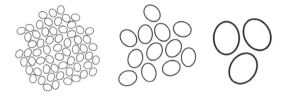

Many small, a moderate number of medium-sized, few large. And this information is not surprising to us, it's a pattern that we have all at least subconsciously observed.

Now – and this is the bit that I find revelatory – if we measure the size of cities around the world, we get exactly the same result:

many small-sized cities, a moderate number of medium-sized cities, and a few very large cities. Exactly the same line of frequency – not just the same general trend, but *exactly the same line of frequency*. Furthermore, if we chart through history the outbreak of wars, it is again exactly the same ratio: many small battles, a moderate number of medium-scale wars, a very few large-magnitude 'world wars'. And furthermore, the stock market exhibits precisely the same pattern of behaviour – the stock market which to me seems a completely false and fabricated reality, yet consisting of and ruled by the same basic building blocks as all matter, and therefore subject to the same governing rules as the rest of the world.

I can't really quite believe this information. That all natural phenomena, including things which seem so inherently inorganic as the stock market, follow exactly the same patterns.

And how do we know this? Because a group of scientists have spent the last decade or so dropping grains of sand into a pile and charting the results. They've tried all sorts of sand of varying sizes: they've tried it wet and dry, they've tried small pebbles, grains of wheat, sugar, all sorts. And they've discovered something called:

The sandpile phenomenon

It's like this: as you continuously drop single grains of sand onto a pile, the pile eventually works itself into a state of criticality, a peak; it forms a complex and interconnected state which is ripe for change. Once this critical state is reached (and natural systems will always work themselves into a critical state), then each additional grain that is dropped has the potential to trigger off a landslide. And we can never predict exactly which grain will trigger it off, nor exactly how big the landslide will be. But we can say that there will be this particular small proportion of large-scale avalanches, this moderate number of medium slides, and this high number of small avalanches.

So, if we can predict generally the trend, why can't we tell exactly which grain will trigger off a landslide? Because of the tiny differences between the placements of the grains – although to our eye one towering sandpile may look quite similar to another, in fact each structure is so unique and interrelated that we can not quite see exactly how the grains of sand affect each other, nor can we predict how one small movement may trigger a massive change on the other side. It's not like a stack of evenly placed bricks, it's like an unstable pile of rocks. No one knew that Rosa Parks's action on the bus on 1 December 1 1955 would set off the chain of events that it did. We can't see how all the individual balances affect each other. There are workings that appear invisible to us, mysterious and mystical, even.

You may or may not find this concept amazing; in some ways it seems startlingly obvious. It's simultaneously incredible and totally credible, because at some level of our consciousness we already instinctively know this. We have seen and comprehended this concept from childhood: that systems are not simple, straightforward structures where individual components behave in isolation. Rather, natural systems are highly complex and interdependent, with many factors influencing each other, and changing any one of the parts may have far-reaching and unpredictable results for the entire system. This is the theory of **complexity**, and whether or not this came about by some grand design or the evolution of chance, accident and the sheer force of life: everything is connected in life, some things in ways we can consciously sense or trace, and others in ways we can't.

This is not a new concept in any way; most religions and cultures have understood this for centuries. For example, let's look at the metaphor of 1 Corinthians 12.

> For the body is one and has many members. If the foot should say, 'Because I am not a hand, I am not of the body,' is it therefore not of the body? And if the ear should say, 'Because I am not an eye, I am not of the body,' is it therefore not of the body? If the whole body *were* an eye, where *would be* the hearing? If the whole *were* hearing, where *would be* the smelling? God has set each of the members in the body just as He pleased. And the eye cannot say to the hand, 'I have no need of you'; nor again the head to the feet, 'I have no need of you.' Much rather, those members of the body which seem to be weaker are necessary. God composed the body that there should be no schism in it, but that the members should have the same care for one another. And if one member suffers, all the members suffer with it; or if one member is honoured, all the members rejoice with it.

If one is to accept and embrace this complexity, this inter-connectedness, if one is to view the world as one whole made from many different parts, then what are we to make of the mechanism of war? In what circumstance would a body choose to wage war on one of its own members? Why, in a healthy whole, would one part, a stronger part, a more arrogant part, perhaps, turn around to another part that seemed, for example, weird or evil, and say, 'It is better for us to eliminate you.' Why would the large and powerful quad muscle say to the foot, for example, 'You do not look right. You have five pointy bits on top and you touch the ground all the time, which is dirty. I am going to kill you.'

Furthermore, who rules? The eye can clearly see, does that make it the ruler? The legs can physically locate the body – does that make them the rulers? Who rules? Those with the power? Those with the vision? Those who can create new life? Who has the wisdom to be able to judge which parts of the body are no longer functioning correctly? Because the knee cannot hear, is it less important? Because the ear cannot bend, is it any less important? The eye is best at seeing. The ear is best at hearing. The leg is best at running. All parts have their places, their strengths and weaknesses, and all parts need each other for the body to live at its greatest possible potential.

We can clearly see in the animal kingdom how this works, and we are all familiar with the interdependence of different species, the potentially devastating effects on the natural order if, for example, we remove even one tiny flea from the food chain. Each animal's well-being is inextricably linked to that of every other. But it becomes more difficult for us to see this in a human society; for example, a local Wellington newspaper recently reported on the homeless drunks who clutter our streets. But what is their role in the web of life? What would happen if we simply 'got rid of them'? Might we find that they have an integral part in the system that we are as yet unable to define?

It was not so many decades ago that one famous short man decided there were certain groups in society who were less useful than others, and began slowly and methodically eradicating them. The categorical wrongness of these actions is difficult to dispute, although it creates an uncomfortable paradox which I will come to shortly. The fixation on 'useful contributions to society' is chilling. Who decides what is useful? Which of us can predict where genius may spring from? Einstein was an extremely slow child, he learned to speak at a much later age than average, and his parents feared that he was mentally retarded. In Hitler's Germany such children would definitely be on the non-useful list. Einstein as we know was vigorously opposed to war.

Let's turn to the poetry of Robert Frost. In 'Mending Wall' he describes two neighbours repairing the stone wall separating their property, which has, by the process of entropy, naturally deteriorated.

> Something there is that doesn't love a wall
> That sends the frozen-ground-swell under it,
> and spills the upper boulders in the sun;
> And makes gaps even two can pass abreast.
> The gaps I mean,
> No one has seen them made or heard them made,
> But at spring mending-time we find them there.
> I let my neighbour know beyond the hill;
> And on a day we meet to walk the line
> And set the wall between us as we go.
> To each the boulders that have fallen to each.
> And some are loaves and some so nearly balls
> We have to use a spell to make them balance;
> 'Stay where you are until our backs are turned!'
> We wear our fingers rough with handling them.
> There where it is we do not need the wall:
> He is all pine and I am apple orchard.
> My apple trees will never get across
> And eat the cones under his pines, I tell him.
> He only says, 'Good fences make good neighbours.

We build these fences around the differentiations of age, religion, gender, 'nationality', physical ability, sexuality, and even dress or hairstyle. The fences are often reminiscent of Dr Zeuss's star-bellied sneetches and those without stars. What difference does it really make? Of course groups will form, and distinctions will be made, but, as Frost says, 'There where it is we do not need the wall: He is all pine and I am apple orchard.' The eye is clearly not the hand, do we need to put a line at the end of the wrist and say, 'Only hands here!'? Furthermore, do we need to ascertain that one is preferable to the other? There are two ways to keep cows in a field: one is a fence, and the other is a watering hole. Fences and walls unleash a destructive instinct in my psyche. It is not my experience that walls and fences work, my testing disproves this hypothesis. When you put a wall in the body, you get a clot. Blood gathers together in a thickened lump, which will then move fatally through the body. This is my experience.

There are two choices when coming across difference: to understand and reconcile, or to judge and eliminate. One is the way of acceptance and embrace, the other is the way of the fence, the way of judgement and rejection. Jack Spratt and his wife had a good system going. Destiny Church[5] does not. If we are to accept the interconnectedness of our lives, then surely good fences do not make good neighbours. They create jealousy, ignorance, suspicion and ill will, and these feelings have a snowball effect that will eventually lead to outright violence and eventually to war.

A dilemma

And here I arrive at a difficulty. If I resolve to never wall anything out, how can I condemn violence? If I am arguing for acceptance of all parts of a community, then how can I judge Hitler's actions? Is he not just another part of the body doing what he is naturally inclined to do? If I am insisting that all natural phenomena be allowed to occur, then how can I argue against the presence of war, which history unmistakably identifies as an ongoing natural occurrence?

Scientists and artists are often keen to avoid ethical questions such as these. In fact, humanity in general displays a frightening lack of responsibility for our actions (we have only to look at the earth's current environmental state, for example). Most religious leaders are only too happy to preach on ethical behaviour, but unfortunately many major religions have been hijacked by angry fundamentalists. Religious movements, in essence if not in practice, strive for a world

5 The fundamentalist Christian movement founded in New Zealand by self-anointed Bishop Brian Tamaki.

in which all people are free and healthy in the fullest sense of the word. However, the attempt to carve out a definition of 'freedom and health' carries an implicit judgement about what is 'not good': that which is 'sinful' or 'bad'. When I asked a Buddhist monk how he would respond to someone whose actions create great harm, he said we should embrace all actions that come from someone in their 'right mind'. Hitler was clearly not in his when he was embarked on grand-scale genocide. The monk used different, more accessible terminology than 'sin', but he was speaking of similar concepts to Christianity.

Religions, though varying in terminology and degree, identify human behaviour which is 'preferable' for the health of the whole. While this works in theory, in human practice it has often become the source of major conflicts and divisions, and there are clearly some faiths which lend themselves more easily to this than others. If there is some sort of 'higher' path, if there is a state of mind which is 'right', then there must be a state which is 'less right', and it is this unfortunately abused differentiation that allows some religious groups to justify their bloody acts of warfare as 'God's will', despite these actions contravening the major establishing principle of their faith – that all are one, that we must love our neighbours as ourselves, and that harm to one part of the body is harm to all.

These arguments also carry personal repercussions. As an artist one often encounters opposition through negative reviews, lack of funding or the unpopularity of one's work. We may wonder if we are in our 'right minds' and struggle even to articulate what 'art' is. Why would we believe so fervently in the importance of our work when the worldly response to our endeavour indicates otherwise? As a friend of mine in the business world says, 'I would suggest there is a very good reason why some artists are not making money from their work.' If financial recompense is seen as the benchmark of value in this society, then teachers, artists, nurses and many others should drastically re-think their existence.

Paul G. Hewitt in *Conceptual Physics* says: 'Both scientists and artists look for the connections in nature that were always there but are not yet put together in our thinking.' Einstein also referred to both practitioners in the same breath, although our society prefers to carve a clear line between the two. Artists are perceived by many as dwelling in some vague, tangential reality that has little relevance to daily life, whereas scientists are seen as carrying out important 'serious' research for the functional world. Products we buy are scientifically proven, not artistically tested.

The artistic process may seem chaotic to others, but there is method to the madness, whether detectable or not to outside observers. Pluto feels the gravitational pull not only of the sun but also of other large, undiscovered bodies, resulting in an orbit that may seem 'wonky' from the majority perspective. To those who feel these alternate gravitational pulls, there is nothing mad nor random about their orbits. They are simply responding to the very real workings of their own personal gravity, some of which is able to be seen and charted, and some of which is not. (There are also the unavoidable black holes into which all of us fall from time to time, unfortunately knowing little about them nor how to get out.)

A scientist said to me, 'Good science is all about the proof. There's no proof for good art – you just throw pots of paint at a canvas and someone calls it genius.' But perhaps the workings of art are like Stephen Hawking's mysterious dark matter: 'clusters [which exert] a gravitational pull on the ordinary matter'. Could this be a test of art, a proof: its gravitational pull? Does it cause a stir? Does it exert a force on the ordinary matter, do the words, musical notations, colourful marks made on a canvas continue to hold an influence over people as the years pass?

Furthermore, at what point can we measure this influence? Within a year? Ten years? A hundred? Learning from history that many artists' work has not been recognised in their lifetime, and that the innovations of such revolutionary scientists as Copernicus, Newton and Darwin have taken decades to be endorsed by their communities, given that Socrates was put to death for his thoughts and behaviour, we must remember that immediate societal approval has never been, is not now and, if we extrapolate, never will be an accurate measure of the importance and truth of a discovery nor of the weight and profundity of a creative expression.

Liberating the pigeons

Frank J. Sulloway says in his book *Born to Rebel*: 'The initiators of scientific revolutions are never scientific communities. The initiators – unconventional people like Darwin – are always individuals with a vision. Some people are inclined to challenge established truths.' We all have our places in the sandpile. The individual grain of sand that causes the greatest landslide does not intend to do so. It simply follows its path, obeying the physical laws of our existence and the conditions of the particular pile on which it was dropped – plus a certain amount of individual unpredictability (the effect of the uncertainty principle). There is no such thing as the right path, only a generalised 'right-ish' area.

We have to embrace the uncertainty and not be paralysed by it, while allowing ourselves to be driven by our wont to categorise and understand. Nature abhors a cage, and yet our inclination is to build one. Even Einstein in his infinite wisdom and acceptance of the great cosmic mystery was hard at work on his 'Theory of Everything', which remained incomplete, unfinished at his time of death. Nature implants in us a desperate need to explain, while our human condition dictates that we will never achieve a comprehensive theory. Further information simply keeps coming to hand.

When Albert Michelson said in 1894, 'It seems probable that most of the grand underlying principles have been firmly established,' he thought that we, as a race, *knew*. Einstein said, 'The most beautiful experience we can have is the mysterious. It is the fundamental emotion which stands at the cradle of true art and true science. Whoever does not know it and can no longer wonder, no longer marvel, is as good as dead and his eyes are dimmed.' He accepted that we didn't know. The truth is that we *sort of know*, we know some of it. It is only at our deaths that 'we shall see in full what we now see in part, then we shall see face to face what is now seen through a glass darkly.' For now, we must be content with shadow-watching, with imaginings and inaccurate renditions of the ultimate source.

As in Plato's ideals, the perfect explanation for the universe exists in theory, in concept, somewhere out there, and will come to us piecemeal, via dreams, intuitions, hard work and accident. But the realisation of this concept will always leave us wanting – to actualise something is to render it imperfect. Some may remain inside their head in their perfect and undisturbed world, yet others will seek to make connections despite the inherently flawed process of doing so. Those who are not afraid of failure, of being wrong, of human weakness and aberrance, these are the individuals who will risk forging the paths ahead, and yet their presence is no more essential than that of those who vehemently resist this change.

And these are the paradoxes that we have to live with – to attempt to know whilst accepting we never can, to be the strongest and most singular individual while knowing we are part of an intrinsically interconnected whole, to delight in the power of speech and discernment while accepting that language is only a temporary and arbitrary measure which divides the One into Many. These distinctions we make are merely lines in the sand, and all at the end is dust, the sun shines alike on the just and the unjust. Yet we will continue to draw the lines.

George Eliot says, 'There is silence enough beyond the grave. This is the world of light and speech.' However imperfect our attempts may be, yet as artists and scientists we must continue to express, to venture into the unknown abyss, to forage for scraps of knowledge, to build bridges between the distant outposts and the thriving central settlements – and to report our findings however vague, inconsequential or downright wrong they may seem; we must place our stones in the pond without knowing if they will fall to the bottom or provide valuable stepping points for others. We must trust that our actions will contribute in some way to the general ongoing movement of the universe and that God is working his purpose out, that life indeed is going somewhere.[6] We must have faith that together all the parts will miraculously come together to make a whole, and that each of us represents a vital part in that, however unlikely and impossible that may seem at the present time.

And thus, with this piece of writing, I cast my stone into the pond and now step back, hopefully awaiting a response. And here I leave you with two final quotations:

---

6 Even if that eventuation is the annihilation of humanity which is highly possible – life will continue somehow, someplace!

There's a very interesting scientific insight which says that regions where real novelty occurs – where really new things happen that you haven't seen before – are always regions which are at the edge of chaos. They are regions where cloudiness and clearness, order and disorder, interlace each other. If you're too much on the orderly side of that borderline, everything is so rigid that nothing really new happens: you just get rearrangements. If you're too far on the haphazard side, nothing persists: everything just falls apart. It's in these ambiguous areas, where order and disorder interlace . . . where the real action is.'

—John Polkinghorne[7]

And from Philip Guston:

Marvellous artists are made of elements which cannot be identified. The alchemy is complete. Their work is strange and will never become familiar.

7   Quantum physicist and Anglican priest, from a *Speaking of Faith* radio broadcast: 'Quarks + Creation'.

Gillian Wearing, 'Signs that say what you want them to say and not Signs that say what someone else wants you to say EVERYTHING IS CONNECTED IN LIFE THE POINT IS TO KNOW IT AND TO UNDERSTAND IT'
c-type colour print, 122 x 92 cm, edition of 10, + 1 AP, 1992–3
(courtesy Maureen Paley, London)

THE PHYSICS ENGINE

by DYLAN HORROCKS

THE
NEW WORLD
BEGAN ON A
MONDAY.

HE WAS ON HIS WAY TO WORK, THINKING ABOUT WHAT AMY HAD SAID.

WHEN SUDDENLY—

I WANT TO MAKE A NEW WORLD.

START FRESH — AND THIS TIME DO IT *RIGHT*, WITH ROOM FOR EVERYTHING...

ALL OF IT BEAUTIFUL

AND REAL

THERE'D BE A FOREST, VAST AND UNTAMED
...
VALLEYS, WITH CARPETS OF PURE WHITE FLOWERS
...
CITIES LOST BENEATH THE OCEAN WAVES
...

AND, ABOVE ALL...

HE DECIDED TO TELL NO-ONE, NOT UNTIL IT WAS READY TO PLAY.

IT'LL BE A SECRET. MY OWN PRIVATE UNIVERSE...

BUT OF COURSE THAT DIDN'T LAST LONG...

I'M THINKING OF STARTING A NEW CAMPAIGN.

OH? WHAT SYSTEM WILL YOU USE?

I-I DON'T KNOW YET. I HADN'T THOUGHT ABOUT IT...

MAYBE D20 - SOMETHING SIMPLE, I GUESS...

HUH, IS IT FANTASY THEN?

FIVE-POINT FUDGE - SET IN POST-APOCALYPTIC TWENTY-THIRD CENTURY HAMILTON...

FRIDAY IS DAVE'S GAME.

WHEN HALF THE PARTY IS CAPTURED BY OIL-HUNGRY CYBERNETIC MUTANT CAT-PEOPLE, ED AND GARY GO MAKE SOME TEA...

HOW'S YOUR NEW GAME GOING? FOUND A SYSTEM YET?

NO, NOT YET. I TOOK A LOOK AT GURPS, BUT I DON'T KNOW...

YOU SHOULD TALK TO PETER. HE'S BEEN WORKING ON A SYSTEM THAT SOUNDS PRETTY COOL...

OH, WELL, I MEAN, Y'KNOW... A HOMEBREW?

NO, REALLY, I'LL BET IT'S WORTH A LOOK. YOU KNOW HE'S WRITTEN FOR MONGOOSE?

YOU'RE KIDDING, PETE?

SURE. HE DID AN AVIATION VEHICLES BOOK FOR D20 MODERN. THE GUY'S A PROFESSIONAL SCIENTIST, SO HE KNOWS WHAT HE'S TALKING ABOUT...

SHIT. I THOUGHT HE WORKED FOR THE MET SERVICE?

HE DOES. I THINK HIS PhD WAS ON THE PHYSICS OF CLOUDS OR SOMETHING...

BUT ANYWAY, THAT'S HIS THING - SCIENCE. AND THIS SYSTEM HE'S WORKING ON IS, LIKE, REALLY WELL RESEARCHED AND EVERYTHING...

THE BASIC IDEA IS TO DESIGN THE MOST REALISTIC SYSTEM POSSIBLE - ALL CAREFULLY BASED ON ACTUAL REAL WORLD PHYSICS, BIOLOGY, GEOLOGY, CHEMISTRY...

SOUNDS COMPLICATED...

YEAH, BUT IF ANYONE CAN DO IT, PETE CAN.

HE'S GOT THIS WHOLE THEORY ABOUT ROLEPLAYING AND FICTION AND SCIENCE...

HE SAYS ROLEPLAYING'S NOT ABOUT STORYTELLING- IT'S ABOUT VIRTUAL REALITY...

IT'S ABOUT CREATING AN ALTERNATE UNIVERSE, THAT'S AS REAL AS YOU CAN MAKE IT.

WOW. PETE SAID THAT?

SURE-SEE? IT'S JUST LIKE WHAT YOU WERE SAYING ABOUT YOUR NEW CAMPAIGN. YOU GUYS SHOULD SO TEAM UP ON THIS STUFF!

YOU KNOW HOW DAVE'S ALWAYS GOING ON ABOUT THE IMPORTANCE OF STORY? HOW HE WANTS HIS GAMES TO FEEL LIKE A NOVEL OR FILM, WITH A WELL-STRUCTURED PLOT AND PACING AND UNDERLYING THEMES AND STUFF?

HUH. RIGHT.

WELL, PETE SAYS THAT REAL LIFE ISN'T LIKE A STORY: "THERE'S NO METAPLOT TO REALITY..."

BUT THERE ARE RULES—LIKE THE LAWS OF PHYSICS, FOR EXAMPLE. AND THOSE RULES CAN'T BE BROKEN OR BENT. THERE'S NO FUDGING DICE OR FATE POINTS OR GM FIAT...

AND THAT'S WHERE PETE'S SYSTEM COMES IN. IT'S LIKE THE UNDERLYING RULES THAT GOVERN REALITY.

I'M TELLING YOU—PETE'S SYSTEM'S GONNA TOTALLY ROCK. PROBABLY SELL IT TO WIZARDS OF THE COAST AND MAKE MILLIONS!

IT'LL BE LIKE THE ULTIMATE PHYSICS ENGINE...

SEE YOU LATER, GUYS!

UM – HEY, PETE? GARY WAS TELLING ME ABOUT YOUR SYSTEM. AND I – I REALLY LIKED THE SOUND OF IT...

I WAS WONDERING IF I COULD TAKE A LOOK AT IT SOME TIME? MAYBE HELP WITH SOME PLAYTESTING?

WELL, IT'S NOT REALLY FINISHED YET. I DON'T KNOW IF IT'LL EVER BE...

SEE – IT'S JUST THAT I – I'M WORKING ON A NEW CAMPAIGN SETTING – AND I KIND OF WANT TO MAKE IT AS *REAL* AS POSSIBLE – SO FOR THE PLAYERS – AND FOR ME – IT'S LIKE ACTUALLY *BEING* THERE...

THAT NIGHT,
HE DREAMED
OF AMY...

NOTES

PAGE 2:

'READY TO PLAY': i.e. TO USE IN A
ROLE-PLAYING GAME. ROLE-
PLAYING GAMES (RPGs) FIRST
EVOLVED OUT OF TABLETOP
WARGAMING IN THE LATE 1960s.
THE FIRST PUBLISHED RPG WAS
DUNGEONS & DRAGONS (1974).
ONE PLAYER (THE GAME MASTER,
OR GM) OVERSEES THE GAME,
WHILE THE OTHER PLAYERS EACH
CREATE A CHARACTER. AS THE
GAME UNFOLDS, THE GM
DESCRIBES THE SITUATION IN
WHICH THE PLAYERS' CHARACTERS
FIND THEMSELVES. EACH PLAYER
DECIDES HOW HIS OR HER
CHARACTER REACTS AND WHAT
THEY WILL TRY TO DO. THE
RESULT IS A KIND OF IMPROVISED
COLLABORATIVE STORY, AS THE
PLAYERS INTERACT WITH THE
GM'S IMAGINARY WORLD.

'...A NEW CAMPAIGN': AMONG
ROLE-PLAYING GAMERS (RPGers),
A CAMPAIGN IS AN ONGOING
GAME, PLAYED OUT OVER A
SERIES OF INDIVIDUAL SESSIONS,
SOMETIMES LASTING SEVERAL
YEARS. THE TERM IS A LEGACY
OF RPGs' WARGAMING ORIGINS.

'WHAT SYSTEM WILL YOU USE?':
A ROLE-PLAYING GAME SYSTEM
IS THE SET OF RULES THAT
GOVERNS CHARACTER CREATION
AND THE WAY IN WHICH EVENTS
AND ACTIONS PLAY OUT. FOR
EXAMPLE, IF A CHARACTER
ATTEMPTS TO LEAP FROM A
FIRST FLOOR BALCONY ON TO
THE ROOF OF A PASSING
CARRIAGE, THE SYSTEM WILL
PROVIDE A METHOD FOR
DETERMINING WHETHER SHE IS
SUCCESSFUL (AND, IF NOT, THE
CONSEQUENCES). MOST, BUT
NOT ALL, RPG SYSTEMS USE
DICE TO RESOLVE SITUATIONS
WHERE PROBABILITY PLAYS A
PART.

'MAYBE D20': D20 IS AN RPG
SYSTEM FIRST DESIGNED FOR
THE THIRD EDITION OF
DUNGEONS & DRAGONS (2000)
AND SUBSEQUENTLY USED FOR

A NUMBER OF POPULAR GAMES,
INCLUDING *D20 MODERN*, *STAR
WARS D20* AND *SPYCRAFT*. D20
IS NAMED FOR THE 20-SIDED
DICE USED TO DETERMINE THE
OUTCOME OF MOST ACTIONS.

PAGE 3:

'THE USUAL D&D BULLSHIT': D&D
IS SHORTHAND FOR *DUNGEONS
& DRAGONS*, FIRST DESIGNED
BY GARY GYGAX AND DAVE
ARNESON IN 1974 AND STILL
THE MOST POPULAR RPG, WITH
MILLIONS OF PLAYERS WORLD-
WIDE. THE MOST RECENT
EDITION (3.5) WAS DESIGNED
BY MONTE COOK, SKIP WILLIAMS
AND JONATHAN TWEET (AMONG
OTHERS) AND WAS PUBLISHED
IN 2002. MANY OF THE MOST
WIDELY RECOGNISED TROPES
OF RPGs (AND SUBSEQUENTLY
COMPUTER GAMES) COME FROM
D&D, INCLUDING HIT POINTS,
CHARACTER CLASSES AND LEVELS,
DUNGEON-BASED ADVENTURES
AND ICONIC MONSTERS SUCH AS
THE BEHOLDER, THE MIND-
FLAYER AND DROW.

'YOU SHOULD USE GURPS': THE
GENERIC UNIVERSAL ROLE-
PLAYING SYSTEM (GURPS) WAS
FIRST DESIGNED BY STEVE
JACKSON IN 1986. IT IS NOW IN
ITS 4TH EDITION (PUBLISHED IN
2004), BUT THE ESSENTIAL
SYSTEM REMAINS THE SAME
(POINT-BUY CLASSLESS AND
LEVEL-LESS CHARACTER
GENERATION; RELATIVELY
'REALISTIC' RULES; THE USE OF
THREE SIX-SIDED DICE, WHICH
GIVES A BELL-SHAPED
PROBABILITY CURVE, IN
CONTRAST TO D20'S FLAT LINE).

'WITH ALL THE OPTIONAL RULES':
GURPS INCLUDES MANY OPTIONAL
RULES, WHICH CAN BE USED TO
ADD MORE DETAIL AND REALISM.

'IT'S A LITTLE CRUNCHY': RPGers
DISTINGUISH BETWEEN 'CRUNCH'
(THE ACTUAL RULES SYSTEM)
AND 'FLUFF' (THE NON-RULES-
BASED ELEMENTS OF AN RPG

GAME WORLD, SUCH AS HISTORY, CULTURES, SCENERY, ETC). A 'CRUNCHY' - OR 'RULES HEAVY' - SYSTEM IS ONE WHICH IS PARTICULARLY COMPLEX AND DETAILED.

PAGE 5:

'5-POINT FUDGE': FUDGE (FREE-FORM UNIVERSAL DO-IT-YOURSELF GAMING ENGINE) IS A RULES - LIGHT RPG SYSTEM DESIGNED IN 1992 BY STEFFAN O'SULLIVAN AND DISTRIBUTED FREELY ON THE INTERNET. WHERE MOST SYSTEMS USE NUMBERS TO DEFINE A CHARACTER'S ABILITIES (e.g. STRENGTH 18, CLIMB SKILL +4), FUDGE USES ADJECTIVES (e.g. 'GREAT' STRENGTH, 'MEDIOCRE' CLIMBER). THE ONLY DICE USED ARE SPECIAL FUDGE DICE, WHICH BEAR NO NUMBERS, BUT INSTEAD GIVE A RESULT OF '+', '-' OR 'O'. CONSEQUENTLY, FUDGE IS POPULAR WITH RPGers WHO WANT TO AVOID NUMBER-CRUNCHING AND FOCUS INSTEAD ON CHARACTER, STORY AND IMMERSION IN THE 'GAME REALITY'. '5-POINT FUDGE' IS A WIDELY-USED CHARACTER-GENERATION SYSTEM FOR FUDGE.

'WHEN HALF THE PARTY...': A PARTY IS A GROUP OF PLAYER-CHARACTERS, AS IN 'A PARTY OF EXPLORERS OR ADVENTURERS.'

'... A HOMEBREW?': A HOMEBREW SYSTEM IS ONE THAT HAS BEEN DESIGNED BY AN INDIVIDUAL PLAYER FOR THEIR OWN USE, AND USUALLY UNPUBLISHED.

PAGE 6:

'WRITTEN FOR MONGOOSE': MONGOOSE PUBLISHING IS A UK-BASED RPG PUBLISHER.

'D20 MODERN': A VARIANT ON D20 DESIGNED FOR USE IN MODERN, RATHER THAN FANTASY, SETTINGS.

'WORKED FOR THE MET SERVICE': METSERVICE IS NEW ZEALAND'S MAIN METEOROLICAL ORGANISATION, PROVIDING FORECASTS AND OTHER INFORMATION TO THE GOVERNMENT, PRIVATE COMPANIES AND THE PUBLIC.

PAGE 8:

'METAPLOT': IN RPG TERMINOLOGY, 'METAPLOT' REFERS TO THE OVER-ARCHING NARRATIVE PLOT THAT BINDS A CAMPAIGN (OR SERIES OF ADVENTURES) TOGETHER. WHEN USED IN REFERENCE TO COMMERCIALLY-PUBLISHED GAME WORLDS (SUCH AS *THE FORGOTTEN REALMS* OR *DRAGONLANCE*), IT IS THE EVOLVING STORY THAT SHAPES THAT WORLD OVER TIME, WHICH GAMERS MAY TRY TO REFLECT IN THEIR INDIVIDUAL CAMPAIGNS.

'THERE'S NO FUDGING DICE': WHEN A GM 'FUDGES THE DICE', HE OR SHE IGNORES A DIE-ROLL RESULT WHICH WOULD 'DERAIL THE PLOT' OR OTHERWISE INTERFERE WITH A SATISFYING GAME EXPERIENCE FOR THE PLAYERS.

'FATE POINTS': SOME RPG SYSTEMS ALLOW PLAYERS TO 'SPEND' POINTS IN ORDER TO CIRCUMVENT THE GAME'S NORMAL MECHANICS - e.g. TO AVOID DEATH IN A SITUATION THAT WOULD OTHERWISE BE FATAL.

'GM FIAT': WHEN A GM ARBITRARILY DECREES A PARTICULAR OUTCOME OR EVENT WITH NO REFERENCE TO THE RULE SYSTEM OR DICE.

'WIZARDS OF THE COAST': THE PUBLISHERS OF *DUNGEONS & DRAGONS*, *D20 MODERN* AND OTHER POPULAR RPGs. TODAY THEY ARE A FULLY-OWNED SUBSIDIARY OF THE TOY GIANT HASBRO, INC.

'THE ULTIMATE PHYSICS ENGINE': A 'PHYSICS ENGINE' IS AN RPG RULE SYSTEM DESIGNED TO SIMULATE REAL-WORLD PHYSICS. THE TERM IS BORROWED FROM DISCUSSIONS OF COMPUTER GAME SOFTWARE DESIGN.

PAGE 9:

'PLAYTESTING': PLAYING A GAME IN ORDER TO FIND FLAWS IN THE SYSTEM BEFORE IT IS PUBLISHED.

PAGE 10:

'MAGIC SYSTEM': IN A FANTASY RPG, THE RULES THAT DESCRIBE AND DETERMINE 'HOW MAGIC WORKS'.

ARE ANGELS OK?
AN ESSAY

CHRIS PRICE

Strenuous intellectual work, and looking at God's nature are the reconciling, fortifying, yet relentlessly strict angels that shall lead me through all life's troubles . . . And yet what a peculiar way this is to weather the storms of life – in many a lucid moment I appear to myself as an ostrich who buries his head in the desert sand so as not to perceive the danger . . .

Albert Einstein, 1897

*If I cried out
        who would hear me up there
                among the angelic orders*

Rainer Maria Rilke, *Duino Elegies*

Prologue

It's February, Wellington, and we're dressing for the circus. Tonight the cool white dome of a tent sprung up on the waterfront becomes our small observatory. We lie back with our hands behind our heads to watch the aerial ballet high above attain the heights of unconscious competence. The violinist and the singer, too, are raised into the temporary heavens. This is where time has brought the melancholy family of saltimbanques both Rilke and Picasso sketched a century ago, who worked the squares of Paris – the father and son, the strongman and the little boy and girl, their sad-eyed mother who twists her skirt in her lap and looks away – they've become a travelling company, having progressed the simple science of gravity and momentum to a perfection pitched at glittering international careers. A long way from pennies dropped on a tatty rug, more out of pity than astonishment.

A long way, too, from the poet who took dictation out of the Adriatic gale at Duino to fix his acrobats and angels to the page. *Don't look down*, we want to say, our hearts in our mouths – the drop to the enormous smallness of the world within's as dizzying as the prospect of the far edge of the expanding universe at the moment when it's poised to start the long fall back towards annihilation. But somehow the music holds them up.

After ladders are descended and bows taken we emerge blinking, shading our smiles against the low sun of late-summer dusk. A boy and girl are turning cartwheels on the grass in a dream of tumbling in mid-air, bending rules, jumping right out of the box.

Centuries ago some people thought the earth's magnetic field was made by the wings of angels as they flew around the planet. Today we're inclined to think an angel is the opposite of accuracy – a shimmering mirage that evaporates on close examination, explaining nothing.

We ask ourselves what happened in the darkness: was it the air disturbed by angels we inhaled that made our voices rise in praise, or just the human frame, overreaching itself in unimaginable ways?

I. 1905: Einstein, Rilke, Picasso

i

The year of the immaculate conception:
    a young man with a wife
        and baby on the way,
who'd barely scraped into the job
    of technical officer third class,
        examiner of patents in Bern,
redraws the laws of physics. The relativity paper
    has almost no references,
        and when famous visitors,
travelled from afar to meet him, enquire
    as to the whereabouts of his theoretical
        physics department Albert Einstein points
to the contents of his desk drawer.

    That's legend – but we should
        know better: each bright
idea has its history,
    its scaffolding and aftermath:
        the thinker
has a mother and a sister with
    a sound skull; a first wife (Mileva)
        who's 'a book', their two
sons in whom the uncertain
    future is invested, and then a second
        (Elsa), more of a book-keeper.

    They're inclined to vanish into
      cosmic background radiation
    in the light of the big bang
      from a world-altering equation.

That, by the way,
    was Einstein's kind
        of verse:
a man who liked

to doggerel, but mostly
    when chasing a particularly
pretty cat. So much
    for thinker's poetry:
        the moon and June
as close as he ever got.

ii

'He was a poet,
    and hated the approximate.'
        At Meudon, Rilke begins
his apprenticeship in accuracy –
    six months of answering letters
for Rodin, entranced
    by his Master's voice –
        one ear to the gramophone
of the physical world,
    acquiring its muscular syntax
        in place of airy hymns.
At Chartres he receives
    instruction on the physics
        of large bodies while an angel
with a sundial averts its gaze.
    'As we neared the cathedral . . . a wind . . .
        swept unexpectedly round the corner
where the angel is and pierced us through
    and through, mercilessly sharp
        and cutting. "Oh," I said,
"there's a storm coming up."
    "Mais vous ne savez pas," said the Master, "il y a
        toujours un vent, ce vent-là
autour des grandes Cathédrales."'
    The air, he said, was agitated,
        tormented by their grandeur, falling
from the heights and wandering
    round the building,
        an explanation admittedly
not entirely scientific

but based, at least,
         on observation.

Another morning the poet wakes
     early, to the richness
         of an unknown voice
singing in the garden.
      Sitting up in bed he can see
          nothing, but later, at breakfast
in the kitchen, Mme Rodin
     whispers happily,
          'Monsieur Rodin rose
very early. He went down
     into the garden.
          He was there with his dogs
and his swans
     and was singing,
          singing everywhere out loud . . .'
Rilke would spend his life
     eavesdropping
          on the happiness of others,
but eventually he had to leave
     Meudon or face existence
          as a light breeze,
a small dog yapping
     round the Master's
          monumental flanks.

iii

That year, in Paris, Picasso took to drinking
     with the clowns and acrobats
          in the bar at the Cirque Médrano.
For him it was the brief dusk of something –
     the shift from blue to rose
          a pause for breath before
he set about the hard maths
     of revising space and time
          in an Avignon bordello.

His *Family of Saltimbanques*
    was taken up
        by Frau Hertha Koenig,
so when Rilke, who moved
    across Europe like a migratory
        bird from one high-ceilinged
apartment or castle of a patron
    to another, arrived in her Paris
        pied-à-terre, he found he'd moved
in with these six itinerants,
    whose space he shared for long enough
        that they in turn began to live in him and,
as any ordinary family under the yoke
    of gravity might, suggested that he
        who could speak should fix them
in print, mixing, if he must, a tint
    of himself into the tale
        so they need not bow forever
to the force that pulled them down.

iv

Light, says Dinah Hawken, is the word
    for light. And so, of course,
        is *Licht*, and *lumière*.
At the heart of the matter,
    the untranslatable –
        in this dog-eared
version of the *Elegies*
    that I've had since student days
        Picasso's painting shows
the huge capital D
    that seems to stand
        for existence – except
that D, as any child will tell you,
    does not stand for existence
        in English, but only
in the German *Dastehn*, a word
    whose simple building blocks mean

'standing there'.
That the irreducible necessity
        of the letter D is dictated
                by the consonance between the letter's
shape and the rough disposition
        of the figures in
                Picasso's painting
makes for an apparently
        intractable problem.

        The thing about science,
any physicist will tell you, is whenever you find
        the answer to one question,
                two more appear.
A question of language
        whether like Dinah you want to say
                that light is one,
or prefer to say wave-and-particle,
        revealing the flaw at the heart
                of the metaphor
which insists that light cannot be
        two, and yet it *is*.
                A question, too, of the particular
history of your language,
        the containers it arrived in,
                whose shoulders you're standing on
to get the view.

        Rilke making the nouns and verbs jump
                through hoops to get closer to it.
Einstein and Picasso conjuring possible
        from impossible with little more
                than a shift in perspective.
The poem/theorem gathering all
        its resources to spring into being
                as if from nowhere –
a palm tree in the desert,
        a fig tree pressing sap
                straight into fruit.

II. 1931–1933: Einstein, Chaplin

So, you like to look at stars. Well then:
    suppose you were taking an evening stroll
in the city of angels – this would be February 1933 – and,
    dissatisfied with a distant glimpse, you'd wanted
to get closer to your heroes, you might have tried
    to climb the fence and hover near the open window
of the house on Summit Drive, where you'd have overheard
    some sinuous equations slipping through
to be absorbed into the Pasadena dusk. And if
    you were bold enough to press your nose up
to the glass, you'd have discovered a corner
    of celebrity heaven: Einstein – who'd spent the day
with Tolman at Mt Wilson, profitably considering
    cosmic rays – and Chaplin repeating a well-tested
Mozart experiment with the resonance of wood,
    the vibration of strings and air and the twelve accepted
notes before an audience of one fat lady
    whose profession it was (wrote Einstein)
to make friends with the famous – a woman quite upfront
    enough to walk in the front door instead of skulking
in the bushes like you, who dare only mist
    the pane with your humid breath before retreating.

And if you were that type, you'd probably have jostled
    round the red carpet with the other stargazers for a glimpse
of Chaplin and the Einsteins arriving at the première
    of *City Lights* two years before:
the relativity circus in full swing,
    and Chaplin that day leaning the whole
weight of his fame against the opening door
    of sound, insisting on the body's speechless
happiness a full four years after the first talkie,
    knowing that, no matter what he said when he finally
opened his mouth, his well-travelled idiom must accept
    the borders of a single tongue.

'What does it all mean?'
asks Einstein, emerging breathless from the noise
        and pressure of the crowd. His host's short answer:
'Nothing.' The crowd cheers equally for the man
        it thinks it understands and for the one it doesn't.

Disguised as a tramp, Chaplin could knock
        at any door and gain entry, but you couldn't
see him straight: what lay beneath
        the too-small coat and outsize shoes
hard to pin down as a cat in a box
        and diamond-cold: 'if you split him,
you would find, as with the brilliant,
        no personal source of those charging lights;
they were only the flashings of genius' . . .

Each scene of *City Lights* a dance
neurotically perfected: over 300 takes,
        a firing and rehiring before the blind
flower girl, an amateur, could play her first scene
        in perfect step with him. In the closing
frames, when she can finally see,
        he gives her all he has – the intricate
and tender choreography of a smile.

For Einstein, this brief time spent
        in the gingerbread cottage in Pasadena
glows in the mind 'like Paradise.
        Always sunshine and clear air,
gardens with palms and pepper trees
        and friendly people who smile at one
and ask for autographs.'
        At the Grand Canyon the Hopi bestow
a headdress and a title – 'The Great
        Relative' – and on Broadway, his name
in a song that spoke for
        the ordinary folk, who'd already become
a trifle weary with Mr Einstein's theory,
        a song where time is still

an arrow, and lovers dwell
    at the centre of a universe in which
the fundamental things apply –

    although the Fundamental Things had packed
their bags a quarter century ago, leaving
    their houses intact, and down-to-earth
and empty. Still
    the words insist: *no matter*
what the progress or what may yet
    be proved, the simple facts of life are such
they cannot be removed. You must
    remember this . . .

Both men preferred a tune that let you hear
    yourself think. 'Listen, play, love, revere –
and keep your mouth shut,' was all
    that Einstein, by now asked for his opinion
on everything, would volunteer on Bach.
    But they could not fail to see the century's
lippy weather closing in, or calculate the cost
    of modern times: a week before the soirée
on Summit Drive, the great dictator
    had arrived in the Reichstag and it was clear
there was no way home.

III. 1939–1945

You can shout all you like, he said,
    no one is listening. We'll call him
Johannes, a minor figure in the hierarchy –
    a guard, perhaps, at the camp in the beech woods
on the hill across from Weimar, where no one's
    looking, either, when men and women march
through cobbled streets each morning to their labour.
    No need for words to get his message across
and certainly no sign of angels. Johannes knows
    the fundamental principles. A smile
is just fear with the corners turned up.

    'The scale of the human heart no longer applies
and yet it was once the unit of the earth,
    and of Heaven, and of all heights and depths,'
Rilke had written, sitting out his world war
    as hairdresser to the heroes, the only job for which
his mother and poetry had prepared him. Ah yes,
    the heroes, who *hurtle ahead in advance*
of their own smiles, who're constantly entering
    the changed constellation of their everlasting risk.

As if a man could climb a ladder
    no higher than his roof and somehow
touch the moon. As if a German-Jewish
    immigrant could tell a President what to do.

In fact, by 1939, a letter from Einstein
    has the power to change the world.
When the world writes to Einstein, it
    proposes marriage, asks if gravity
makes a man fall in love, proffers weird
    theories, anti-Semitic rants, and tells him
he should get a better haircut. It addresses
    the envelope to Professor Albert Einstein,
Master Tailor of Clothes for Vacuum Space.

The world goes in the funny file, but
his letter finds its mark; before long, at Los Alamos,
the bomb's on fast-track.

IV. Afterwards, and before

Consciousness, wrote Milosz, *brings no solace,*
       since it is the consciousness of a clown turning
somersaults on a stage, hungry for applause.
       (Don't forget that Milosz, after all
he went through in the war, had ended up
       in California.) Whether or not you agree
may depend on where you live. In
       this morning's *New York Times,* Deborah Solomon
interviewing the British historian of *Ideas:*
       'You strike me as deeply unanalysed.
Have you ever considered seeing a psychiatrist?'
       To which the answer: 'I *was*
a psychiatrist. I left because I thought Freud
       was rubbish.' A viewpoint Einstein shared,
while respecting Freud in public for his eloquence
       and for politeness' sake. His sons stayed loyal
to Mileva, but failed to shake their famous
       father off. Hans Albert stubbornly determined
to work with water, old knowledge needing nothing more
       than Newton and an engineer's idea
of the line of least resistance. But young Tete,
       Eduard, the one who loved words, hung
Freud's photograph above his bed and began
       to study for psychiatry. In 1932
he entered the Burghölzli as a patient.
       'Nothing is worse for man,' he wrote,
'than to meet someone beside whom
       his existence and all his efforts are worthless.'

An angel is a mountain range, a mirror.
       An angel is a hard, bright, icy creature.
The Great Relative prefers to dwell in 'the darkness
       of not-having-been-analysed',
displaces melancholy with music and
       hard work. Max Brod, who played Mozart with him
in a Prague salon, cast him in fiction as

the 'spotless angel', Kepler, who served
not truth but only himself, his own purity
and inviolateness. Worse is,
of course, to come; a year later
Einstein sails for America.
He will not see his younger son again.

Only Elsa knows how, underneath
her husband's eerie surface calm, Tete's illness
pains him. Within the small universe
of his violin, the heart's scale still applies –
sonatas make their clearing and lay down
a carpet where the acrobats of feeling
can perform their tricks unseen.

V. 2005 (and beyond)

In California and Wellington
    telescopes cock an ear
        to the expanding universe.
Emptiness drifts from the heavens
    in great gusts, and with ninety
        per cent of the all of it
currently listed
    missing in action,
        perhaps a poet's ideas
of angels aren't such a stretch:
        joints of pure light, corridors,
staircases, thrones,
    pockets of essence
        or *suddenly separate*
mirrors that gather their beauty back
    into themselves,
        closed systems conserving
energy. There's a growing list
    of impossible things
        to believe before breakfast.
Take those mirrors, for instance:
    it's said (and only half
        in jest) that
if you could transport yourself
    to the photon sphere
        (that's light in orbit
round a black hole)
    you could in theory see the back
        of your own head.

As time goes by it's easy to forget
    the high cost of achieving
        the impossible.
On the edge of the Grand Canyon
    the Hualapai have built
        a glass platform
where even the timid can walk out into air

and look straight down
        through time at the bright
watersnake of river far below.
        Back in Las Vegas, at Bellagio,
            Cirque du Soleil perform
the umpteenth repeat of their water show –
        'O', so much talent diving
            into such a shallow pool.
Somewhere in this desert
        a giant cloud-palm sprung magisterially
            from utter dryness cast its shade
across the remains of the century,
        but today the city of glittering
            fountains, palms and light grows faster
than any other in the world, afloat
        on Colorado snowmelt, grace enslaved
            to power. Buses take the tourists out
past the guards and guns of homeland
        security to see the ring around
            the slowly draining bathtub
of the Hoover Dam

        and tonight, televised live
            for a worldwide audience,
a few grains of comet's tail snatched
        from the beginning
            of the solar system will parachute
into the desert near Salt Lake City, dust
        to dust. Truth in a dry climate often a matter
            of coaxing the implausible to appear
from the wings: palms from piped water,
        angels or demons from a mind deprived
            of context, or deep space dropping from the sky,
its particles captured by a gel almost entirely
        made of air. Not even close
            to fathoming how far we'll go
before we really meet the nothingness
        without, or the vacancy
            within.

VI.

I've spent the summer flitting
    between borrowed houses, enjoying
the patronage of friends. In a Paekakariki bach
    Picasso's in the bathroom and here
in this high-ceilinged North Shore villa
    where my rent's to water plants
and feed the cat, Einstein's on the fridge,
    pinned there by a small
gold plastic angel with a lute.
    Weirdly retouched, his face is
simultaneously young and old, the suit
    implausibly well-pressed. Inscribed across
the bottom of the photo in slanting
    blue letters, his axiom 'imagination
is more important than knowledge' –
    a saw to reassure the most of us
who lack the appetite for hard maths that
    I'm ok and *you're ok* with nothing but
an unquantifiable 'creativity'
    to sustain us – a thought to succour
Sunday painters everywhere. By now
    words and pictures are an old
married couple, unable to imagine life apart –
    already half a century or so since
someone dressed up Chaplin's theme
    for *Modern Times* with a title,
'Smile', and a sugarcoat of lyrics
    to clothe the naked bittersweetness
of the tune in one colour-coordinated
    costume, a top-ten hit for Nat King Cole.

You'll even find our poet on the shelves
    as New Age guru here in *Rilke for the Stressed*
and making guest appearances in books that aim
    to rewire the American corporate brain
with the aid of 'quantum self'. Fact is that he, like Einstein,

rejected indeterminacy, thought the Impressionists
'no better than the realist school,
        painting "I love this"
instead of painting "Here it is"'. Better,
        he thought, the consuming of love
in anonymous work that he found
        in Cézanne.

This villa is a woman's anonymous creation:
        roses, lobelia, grapevine, yucca, pansies,
basil, parsley, alyssum are massed
        on a verandah deep and generous enough
to work outside even when it's raining,
        although it's hardly rained a drop.
In the garden, the phoenix palm rattles its sabres.
        On first doing the watering, I thought: the tyranny
of creating something you can't leave
        or it will die. The next time: the necessary price
for making something beautiful and alive.
        The third: if you have friends
or family they will tend it for you. A syllogism
        in there somewhere.

On the radio, old songs, a summer's
        low-key chatter: this year's plan
to get the road toll down and now
        the council wants to ban the phoenix, fan
and bangalow palms that menace the natives
        with their profligate self-seeding. A doctor
who sees ten kids a year with palm-frond injuries says
        he'd be glad to see them gone. Sometimes
it seems we live in ever-narrowing parameters
        of safety, as if risk
might one day be confined to an existence solely
        in the mind – Einstein cutting his own throat
while shaving in a dream after leaving Mileva.

*

It's true, if you sit and stare for long enough
        the unquiet mind gets up its own entertainment –
a figure arrives to fill the empty stage, a ghost
        or shadow-minister or even, if the need is great
enough, an angel. I'm thinking of Noel Mangan,
        the operatic bass, at the moment he stepped out
on stage to sing and no voice came (the cancer by then
        had its hooks in him) – until he felt a buttery light
and warmth, and arms slipping round him from behind
        to support him so singing could return
full strength until the show was over.

When you're improvising, says the bass player
        from The Necks, don't look down – the mind,
a kind of quantum computer, will perform
        all operations simultaneously, provided
no one's looking. I'm thinking of Rilke at Muzot,
        deep in the disinfected soul of Switzerland,
pulling sonnets from the air for days on end
        after ten years of nothing much but waiting
in which somehow he'd become a juggler
        so practised that the balls dropped to his hand
like falcons to the gauntlet, and then at last
        the remaining Elegies stooped fiercely down and gripped
the gloved and waiting wrist. Sometimes
        you have to turn away so you can see.
Call it grace, if that's what may appear
        after years of the invisible gathering
of speed, coasting with the engine
        in neutral. Truckies call it angel gear.

VII.

This year research has totted up
        the cost of the muse – bottom line
            a greater likelihood
of early death or mental
        dissolution. No real
            surprises there, although
you might have hoped the cliché
        would be undercut by fact.
            Einstein was known
to say that each man measures
        according to his own shoes
            (a cliché, like as not, in German).
Everyone their own
        devastation. Each on its own scale, is how
            Hass puts it at the end of *his*
long argument, *I know we die,*
        and don't know what is at the end –
            Rilke so determined to find out
he refused all help
        to make the ending bearable
            in order to retain himself,
conscious and in agony till the finish,
        the experiment of a lifetime seen through
            to its conclusion.

            A poem may have no reality
test but time – but is it really
        necessary, Rainer Maria Rilke,
            to die unreasonably
young, to opt, as Lowell did when revising
        his literary life, for *the boy Keats*
            spitting blood out for time to breathe
over a place in the index
        of his correspondents?
            Science may do its best work
early – Einstein's five great papers all

written at the age Keats died – but if you'd
                gone before thirty, you'd have left
your best work undone.

                        And if that were so,
            Einstein returns to whisper
                mischievously in my ear,
would the world have been
                worse off? Surely all this fuss
                    with words is just
a flea circus beside the work
                of those who found a cure
                    for Keats' disease?

        *

At Foxton Beach the stilts and oystercatchers thread
                their cries on the wind. On the estuary,
Siberian godwits thrust their beaks up to the hilt
                in mud. Sea sifts and lazily discards the light,
the present moment's revelation.
                What could a poem have to do
with truth? Rilke was proud of 'The Ball'
                because it expressed 'pure movement',
as if the poet had rolled up his carpet,
                leapt into air and vanished
in a puff of smoke. Like other conjuring tricks,
                this one's both true and patently false,
relying on suspended disbelief
                or a manipulation of the point
of view. Or a question of scale, perhaps,
                since water performs
its algorithms oblivious to our assessment
                or opinion, but light
is crucially affected by the means
                of looking. If physics diagnosed poetry
it might take Wolfgang Pauli's
                scathing phrase to say the patient's
not even wrong – a theory so ill-formed

it can't be tested against reality.
                                        OK:
we all live somewhere on the crackpot index.
        But let's suppose, more kindly,
that a poem's anecdotal evidence – one bird singing
        on the tree outside your window
that becomes evidence of dawn
        when it's joined by other birds. Unreliable,
but irresistibly present, a song is a stitch
        in time, a knot securing now in the hope
tomorrow might be taken care of.
                                        And if,
as master builder Milosz also claims,
        its load-bearing strength is such that
One clear stanza can take more weight
        than a whole wagon of elaborate prose,
then maybe a life *can* hang
        from a thread of song.

For the sake
        of the argument then
we might allow the angel as placebo –
        a useful fiction, with a physical effect.
And if the singer is the engine
        of his own transcendence – if a brain,
under the right conditions, can
        manufacture its salvation –
what does it matter, if
        the end's the same?

Epilogue

It's Wellington, early March. Last night on stage we saw *The Bright Abyss*, with Chaplin's barefoot grandson reincarnating his moves and adding another layer all his own. This evening, it's our turn: in a carpeted room above a car dealership on Cambridge Terrace we lie down and learn to rearticulate ourselves muscle by muscle, gesture by infinitely minute gesture. After class we walk downstairs into the summer evening temporarily upright, two springs from which a weight has been released, and our feet seem to trust the earth – we had not known how doubtful they had been.

In the physics lab, when Marcus showed me the experiment demonstrating waves, I recognised its principles from this class, and could tell his students – who were struggling to get a signal on the machine that registered their breathing – where to place the strap to get a better reading. How can we, who scarcely inhabit our bodies with ease, aspire higher than a rooftop – you with your damaged disk and dodgy knee, me with hip pain from years of walking like the little tramp?

Feldenkrais, the engineer who devised this bodywork to fix his own injury, prized reversibility in all things, aiming to make the impossible possible, the possible easy, and the easy elegant. Until Maxwell's imaginary demon steps out of the wings to vanquish entropy, though, there are some things that can never be taken back – not even down the path of least resistance – but only shuffled forward, in whatever way we know. 'A butterfly,' said Einstein, 'is not a mole; but that is not something any butterfly should regret.'

There's a poet in this town who will not learn to dance. He says with a sly grin that if you could film inside his shoes, you'd see his toes enjoying small, astonishing moments there. Like the life-or-death riddle of Schrödinger's cat in a box, we'll never know if this is true, but in the paradox is a lightness of being that stacks up – for now – to something more – or less – than words.

AFTERWORD:
LUMINOUS MOMENTS

PAUL CALLAGHAN

Wanganui has a river crossed by bridges and hills topped by towers. In one of those hills is a tunnel built in 1919, a tunnel with painted arched walls and a repeating line of ceiling lights, striping a path 200 metres into the hill. At the end of the tunnel is an old scissor-trellis door, and a lift that wobbles its way to the hilltop, a lift designed to frighten young boys who have ridden bikes to what seems the centre of the earth. Entering the tunnel, even a slight rustle generates an echo. Noisy boys shout and then wait for the returning voice, more than a second delayed. The experience is magical.

I don't think that as a boy I ever consciously thought about the speed of sound, but I sensed that it had a speed, and that it was slow. Life was full of extraordinary mysteries then. For example, when I was very young, there were the Saturday afternoon movies. My friend Billy thought that there were real cowboys on the stage, and I argued that there weren't, but I didn't quite know how to explain that properly. Dad had a magnifying glass, and my big brother had learned how to make a slide show, with a sheet hanging on the garage wall, and the magnifying glass lens and a light bulb at opposite ends of a wooden apple box with a hole cut by fretsaw that allowed the lens to exactly slot in place. We made our transparencies by drawing frames on a roll of greaseproof paper that we ran through slots on either side of the box and between the bulb and the lens. There seemed to be a special placement of all these parts that made the picture on the screen sharp. Oddly, the transparent paper had to be positioned with the pictures upside down inside the box, if they were to be upright on the screen. All that was a puzzle, but I learned to make it work. It certainly impressed all the neighbourhood kids who came to the pictures at our garage.

I loved that garage with all the tools and paint tins. I nearly burned it down by throwing fireworks inside, but Dad and my uncle Bill managed to make their way through the flames to get the hose, and so only one end of the garage was burned out. Anyway, Dad nearly burned the house down once, by putting old embers from the fire round the back and too close to the wall. Maybe that's why he was so gentle with me then. But when I dug a really deep hole in the garden and pushed the lawn roller with its huge concrete wheel over the edge, Dad was annoyed. To get the roller out he had to excavate a long sloping ramp from the bottom of the hole to the garden surface, I guess using the same method the Egyptians used to build pyramids. He told me then, 'I hope you have a son who causes you as much trouble as you cause me.' Later, he built me a cart with steering ropes. It went really fast down the hill towards the river. We were so lucky having a river in Wanganui. My friends and I built a boat out of corrugated iron caulked with tar taken from the road sealers. It was OK for a while, but then it sank in the river but that didn't matter because we could all swim.

We had a piano in our house and all of us had piano lessons. At age 10 I started to learn to play the cornet, because some of my older cousins were in a brass band and I wanted to try that. It was amazing that I could play different notes without having to push the valves. Those notes seemed to form the same arpeggio as on the piano. Mum's vacuum cleaner hose worked too, but sounded more like a euphonium, and the sharp ends were hell on my lips. The house was full of fascinating things, like the radio with glass bottles that glowed, and Mum's electric sewing machine that once threw me across the room when I stuck my finger in the empty light socket.

By the time I went to secondary school I had done a lot. I had already seen satellites passing over the sky, and I had built model planes, though never, like some of my friends, proper ones with a propeller and a glow plug motor. I had also built a crystal set. We used diagrams and bits and pieces provided by other kids or bought in model shops. The coil had various places where you could connect the crocodile clip and you had to get the right place to connect before you could select a radio station by moving the comb of plates that slid inside each other. That sliding part was called the condenser. I liked that word because it reminded me of condensed milk. I could only find two stations though, one very serious that sometimes had Parliament broadcast on it, and one local station with lots of music, like the Everly Brothers and Elvis Presley. To get these two stations I had to string a long aerial wire across the garden between poles, and also connect an earth wire to a metal pole stuck in the garden outside my window. I had no idea why that was. Radio was a mystery.

At secondary school there were chemistry laboratories, and the bright reddish flames of potassium thrown in water. My friend Richard Green knew where to buy chemicals on mail order using our pooled pocket money. Chemistry was about explosives and astonishing colours and strange smells. But physics and mathematics eventually won me over. The echoes of my childhood had posed too many questions begging answers. And physics gave answers, expressed often in the language of mathematics. It also quantified the natural world. By the time I was 17 years old I had measured both the charge and the mass of an electron, using our teacher's brilliant but rudimentary school equipment, incorporating coils of wire similar to that of my crystal set. Rutherford once said, when asked if the electron existed, 'Why, I can see it as plain as that spoon in front of me'. I think I grasped Rutherford's way of 'seeing' way back then. The reality was in the measurement, in the self-consistency of the ideas behind the interpretation. Science had become more than explosions, or the mysteries of crystal sets and magnifying glasses. It had become about a way of seeing the unseeable, of reaching into a world of imagination founded in both measurement and mathematics.

I knew then that I wanted to study physics. But still I really didn't know the thrill of science, the way that science really takes hold of you at the moment that you begin original research and start to discover for yourself. In many ways the years of university training were an interregnum between the playful experiments of childhood and the real research of my early adulthood. I joined a great research laboratory where I cooled atomic nuclei to within a thousanth of a degree of absolute zero, using their gamma ray emissions to see which way they pointed, and randomising their orientations by resonating them with radio waves. The mundane level of observation, the laboratory connection to it all, was the response of the radiation counter to the sweep of frequency from the radio source. But the reality to me was the complex gyration of the atomic nucleus, a process which I could only grasp through my imagination, and express precisely through mathematics. I cannot explain easily the sensation of that imaginative process, except to say that it thrilled me and placed the everyday life outside the laboratory in an especially happy context. At moments when the creation of some new experiment in my mind brought eventual success in the implementation, the feeling was almost an ecstasy. I remember when I was 36, and I had dreamed up an experiment which, if properly executed using the right radiowave stimulation, would cause subtle nuclear gyrations revealing very slight nuclear shape distortion. It worked perfectly and afterwards I lay awake all night, just reliving the experience, not so much the moment of truth in the lab, but the experience of visualising the atomic nucleus in its dance, of causing it to display its hitherto unseen shape, as well as the electronic environment of the material in which it was immersed.

I think my whole life in science has been uniquely connected to the type of research I do. I have lived in this world of the nucleus, using radiowaves to induce the atomic nucleus to reveal information about its molecular and material environment. To distant relatives and funding agencies I justify this by pointing out all the valuable information I can gather about plastics and food products and plants and people. But the plain truth is, I'm hooked on the world of imagination that the atomic world allows me to enter. I guess if I were a collector of butterflies or a pathfinder for new molecular synthesis, I would be motivated quite differently. I'm quite sure that personality plays a role, that it makes us do the sort of science we do, makes us think the sort of creative thoughts that original science demands. All of us who do science are different personality types. But we all labour under the same discipline. There is always the same tension in the creative process of science.

Richard Feynman once said: 'Our kind of imagination is quite a difficult game. One has to have the imagination to think of something that has never been seen before, never been heard of before. At the same time the thoughts are restricted in a straightjacket, so to speak, limited by the conditions that come from our knowledge of the way nature really is. The problem of creating something which is new, but which is consistent with everything which has been seen before, is one of extreme difficulty.'

In science, there is a brutal process of judgement that cares nothing for the age, or status or brilliance or track record of the practitioner. Science subjects us to the court of observation, where accordance with reality, consistency with all we know, and ability to go out on the limb of new observational challenges and succeed are all that matter. Scientists can be competitive or cooperative, selfish or considerate, boisterously extrovert or quietly introvert. But one thing they all share is the need to admit, on a regular basis, that they got it wrong.

As part of *Are Angels OK?* I spent many hours talking with both Jo Randerson and Vincent O'Sullivan. They are such very different personalities, as different as any pair of scientists one might meet. Vincent quietly absorbed during our conversations, listening more than talking, his eyes sparkling as I took him around my lab and showed him the things we do, the tools we play with. He was fascinated by elastic fluids, the stuff of the spider's web, and he wanted to read in depth on the subject.

Jo is effervescent and talkative, and we shared a torrent of words, ranging far and wide over many aspects of physics. She was really taken with the idea of 'self-organised criticality', where complex systems poise themselves in apparent stability, broken by bursts of upheaval. The behavioural commonality of avalanches on sandpiles, ecosystem evolution and stock-market fluctuations intrigued her. Jo explores the religious dimensions of life, and we talked about how some natural systems can become so complex that physics can't really describe them in any predictive way, and that led to my acknowledgement that science may have its limits. There is a place for God, if you want that, but I preferred not to go too far down that conversational path. The language just gets in the way. When we physicists speak of 'God', we mean 'the way nature works'. It's a private thing, spirituality.

Although Vincent and Jo never discussed their creative lives explicitly, I would guess that they too live with 'constrained creativity', where the intimidating body of existing literature demands a freshness of approach, an awareness of what has gone before, perhaps, but a need to say something new. Remarkably, Jo and Vincent were fascinated by the world of science, and all the other physicists had the same experience in their respective conversations with partnered writers. For example, Jo and I found that we were able to talk about quite complex scientific ideas together, though we had to resort to the 'parables' of everyday phenomena. In physics, one can only go so far without mastering the mathematical language by which nature expresses itself. I think, however, that all the writers and physicists shared much in common, as creative people fascinated by the world around them.

And, of course, we scientists are all writers, or at least we need to be writers if we are to communicate our work effectively to the world of science. The truth is that the work of another scientist often seems just as mysterious to us as it can be to the non-scientist. Great science often requires that we express the most subtle and original ideas in simple, compelling prose. Sometimes we try to break out of the straightjacket of precision. Most scientists long to indulge in metaphor, leaping at the chance to speak of 'colour charge', or to name some elementary particles 'quarks', or to refer to the origin of the universe as 'the big bang'. Language, and its power to entice and fascinate, are central to the world of science, and, indeed, I am sure that many are attracted to physics by the desire to become more familiar with the delicious vocabulary of that discipline. Maybe that is true of other branches of science as well, or maybe geologists just like to wander in the mountains and chip at rocks, while entymologists like to collect insects and marvel at nature's bizarre morphologies. While all of science has lovely words, words that intrigue and fascinate, it's not at the heart, it's not the 'nub' of science, in the way that words and their associated feelings and memories are the 'nub' of writing. In the end, however we communicate, whatever motivates us, whatever lured us into our research specialities, we scientists are all ruled by the harsh requirements of accordance with observation. At that moment of truth our literary leanings, our personal characteristics, whether we are wild and passionate or careful and meticulous, all must be submerged. We have to face facts, and face regular failure. That we all share.

At least we scientists may share our burdens, pool our personal traits, and work in common, drawing on our different strengths. For creative writers, the struggle is usually lonely and peculiarly personal. For scientists, we mostly work in teams, indeed almost always in teams in the case of experimental scientists. And even if, as theorists, we calculate alone, we are motivated to find accordance with the observations of our experimental colleagues. Science is an intensely social activity.

I think that we scientists share with writers the diversity of personality factors and early experiences which determine the almost infinitely varied ways in which we are motivated, in which we create, in which we ply the craft of our work. Those 'echoes of my childhood' resonate in so much that I do. I remain fascinated by the natural world, by its mysteries, by its beautiful phenomena, and by the man-made technologies driven by natural laws. My continual hope is that repeated failures and puzzlement will be punctuated by occasional luminous moments. I think that all the writers here would probably share that hope.

NOTES ON CONTRIBUTORS AND COLLABORATIONS

Particular collaborators are credited in the background notes which follow; however, this book and the broader *Are Angels OK?* project itself would not exist without the knowledge and generosity of the following scientists:

Geoff Austin, University of Auckland

Rob Ballagh, University of Otago

Phil Butler, University of Canterbury

Paul Callaghan, Victoria University of Wellington

Howard Carmichael, University of Auckland

Matthew Collett, University of Auckland

Pablo Etchegoin, Victoria University of Wellington

Richard Hall, Stonehenge Aotearoa

John Harvey, University of Auckland

David Hutchinson, University of Otago

Kate McGrath, Victoria University of Wellington

Tony Signal, Massey University

Alistair Steyn-Ross, University of Waikato

Moira Steyn-Ross, University of Waikato

Jeff Tallon, Victoria University of Wellington

Matt Visser, Victoria University of Wellington

Michael Walker, University of Auckland

Andrew Wilson, University of Otago

David Wiltshire, University of Canterbury

PAUL CALLAGHAN

Paul Callaghan was born in Wanganui, New Zealand, and studied physics at Victoria University of Wellington. After a DPhil degree at Oxford University, he returned to New Zealand in 1974 and took up a position at Massey University, where he began researching the applications of magnetic resonance to the study of soft matter. He is currently Alan MacDiarmid Professor of Physical Sciences at Victoria University of Wellington. Paul is a well-known public speaker on science and regularly talks with Kim Hill on her Saturday Morning radio show. In 2001 he became a Fellow of the Royal Society of London and in 2006 he was appointed a Principal Companion of the New Zealand Order of Merit.

HOWARD CARMICHAEL

Howard Carmichael holds the Dan Walls Chair of Theoretical Physics at the University of Auckland. For *Dead of Night* he translated the first chapter of Genesis into mathematical formulae. He was born in England and moved to New Zealand at the age of two. Educated at the universities of Auckland and Waikato, he travelled to the United States for postdoctoral work in 1977. He held positions as researcher and teacher at various institutions in the US for the next 25 years. He moved to Auckland from the University of Oregon in 2002. He is the author of more than 100 scientific papers and two research monographs in the field of quantum optics.

CATHERINE CHIDGEY

Catherine Chidgey was born in New Zealand in 1970 and grew up in the Hutt Valley. She has degrees in creative writing, psychology and German literature, and lived for two years in Berlin, where she held a DAAD scholarship for postgraduate study. Her much-awarded first novel *In a Fishbone Church* (1998) has been followed by *Golden Deeds* (2000) and *The Transformation* (2003). In 2002 she won the inaugural $60,000 Prize in Modern Letters, while in

2003 she was judged the best New Zealand novelist under 40 in a *Listener* critics' poll. She now lives in Dunedin, where she has recently been Burns Fellow at the University of Otago.

Catherine Chidgey writes:

When I talked to the two Davids [Hutchinson and Wiltshire], I felt like a first-year student – I did a lot of note-scribbling in the hope that it would all make sense later, and I was most surprised when some of it actually did. I began about five different stories, none of which, I felt, had any 'heart' – my biggest problem was to make the leap from theory to a believable, breathing character. When I settled on the (very broad) topic of gravity, however, I realised that I could use many of the terms metaphorically as well as literally, and the story grew from there. Pressure, load, weight, force, how much a person can bear – thinking about the different meanings of these terms told me about my main character's nature and relationships as well as about his special physical talent.

GLENN COLQUHOUN

Glenn Colquhoun is a doctor, poet and children's writer. His first collection, *The Art of Walking Upright,* won the Jessie Mackay Best First Book of Poetry award at the 2000 Montana New Zealand Book Awards. *Playing God*, his third collection, won the poetry section of the same awards in 2003, as well as the Reader's Choice award that year. He has written three children's picture books and published an essay with Four Winds Press entitled *Jumping Ship.* In 2004 he was awarded the Prize in Modern Letters. He works as a GP on the Kapiti Coast.

Glenn Colquhoun writes:

I did not study physics at high school. I took history and geography instead. I was good at maths in the fifth form, but slowly forgot its secret handshakes after that. Science came to me at medical school.

I took a course in remedial physics and failed it. I sat medical physics and failed that as well. I was appalled, not just because I failed but because I loved it. No one else in the class seemed to. It felt unjust that someone who loved physics could be incompetent at it while those who were competent at it didn't care. Physics was the best-looking girl in the room. She only went out with the best-looking boys. I was the guy she should have married. She never knew it. I would have treated her right. I told her she'd be back. And now she is.

The problem was maths. I couldn't do it. I remember writing out each new equation I was introduced to in long sentences so that I could see what it was saying. It was always beautiful. When it came to manipulating them or plugging in data I was hopeless. Once, a lecturer derived the exact energy released from a hydrogen electron when it is released from its orbital. The blackboard was covered in writing. I copied it down like a short story and learnt it by rote. It seemed to me for the first time that physical equations were like poems. They were a form of shorthand for elegant thoughts. I remember sitting in church that week and secretly following his reasoning on a small piece of paper held in my hand. If I looked especially devout, I was. He told me later – when it became clear that as well as being devout I was an idiot – 'No one who fails physics should ever become a doctor.' I know a number of patients who would agree with him.

Given the opportunity to be part of this project, there was really only one thing I wanted to do. I wanted to go back to those equations I used to write out in longhand at medical school. I wanted to apologise for failing them. I wanted to tell them I had kept the faith. I wanted to tell them that they were still some of the most beautiful poems I had ever read.

With the help of Tony Signal at Massey University, I chose 10 well-known equations in physics. I accepted that if mathematics was a language, then these equations were its poems. Tony walked me through the physics, then I went to work translating them into poetry. I need to say a huge thank you to Tony, who has been unfailingly patient with me through this whole process. He reminded me at times of a long-suffering midwife who has seen too many women have babies to think it can't be done this time. Confronted by the Yang-Mills Langrangian he calmly boiled water while I screamed for the epidural. He has been a wonderful guide and a good friend. These poems are his godchildren.

I have to say at the outset that my translations are not explanations. Instead they are attempts at translating the core relationship hinted at in each of the equations. To me poems and equations are both compacted forms of language. They rarely explain or qualify themselves as they go. For this reason they can become threatening. People often make excuses to leave when they see them coming. At best they are witty and exhilarating. At worst they are snobs. When they work well, good poems and good equations comfort and predict and exhilarate.

For me, above all else, both science and art are engaged in the great adventure of hunting for the right metaphor. They are both trying to find the perfect way of explaining something – searching for the exact pattern that will capture or outline those relationships in the world we are desperately seeking to understand. These patterns are the great missing links that reconcile what is unseen or unappreciated with what is ordinary and everyday. We stalk them with sticks and with kisses and with small pieces of bread spaced at just the right interval in the backyard.

With these poems my short career as a physicist ends.

DYLAN HORROCKS

Dylan Horrocks's graphic novel *Hicksville* was the first of its kind to appear in New Zealand, and since 1998 it has been translated into several European languages. His comics won an Eisner Award in 2002 (Talent Deserving of Wider Recognition) and have been nominated for a number of other international awards. He has drawn strips for magazines and books in New Zealand, Australia, England, the USA and France, and has also written comics for DC Comics and Vertigo, including 25 issues of *Hunter: The Age of Magic*, 19 issues of *Batgirl* and three issues of *Legends of the Dark Knight*. He is currently working on a new series for Drawn & Quarterly called *Atlas* and a graphic novel for Top Shelf. He lives in Auckland. His website is at http://www.hicksville.co.nz.

Dylan Horrocks writes:

I avoided science at school. It always seemed to involve smelly chemicals, frighteningly temperamental Bunsen burners and an endless array of opaque formulae. I did like maths, however, but most of all when it had nothing to do with practical applications. It was the sheer pleasure of unravelling intricate, meaningless puzzles I enjoyed.

Since then, I've occasionally read popular science books – on chaos theory or the human genome – and always found them exciting and inspiring. But this project was the first time I've really been obliged to face up to science properly. It proved to be even more fascinating than I had imagined; but a lot more challenging and difficult, too.

I don't mean difficult in the obvious way (trying to get your head around special relativity or light teleportation, for example); that stuff was hard work, to be sure – but fun, too. No, I'm talking about a deeper kind of challenge – as a writer and an artist.

Picasso famously said that art is 'a lie that tells the truth'. Certainly, writers of fiction lie all the time. We tell stories about things that never happened, to people who aren't real, in worlds we've made up. Even supposedly realist novels aren't actually 'real', but take place in an *imagined* reality; one that exists only inside the author's head.

I've been interested in this aspect of fiction for a while, but the more time I spent talking to physicists and reading about science, the more troublesome it became. It seemed to me that at the heart of science is a total commitment to truth, to understanding and explaining what is *real*. The greatest sin a scientist can commit is to fabricate evidence – to *lie* – and claims that it was all in the name of some supposedly greater 'truth' won't cut the mustard (as many a juicy scientific scandal has shown).

It came as no surprise, then, when the physicist Geoff Austin told me he has little patience for fantasy fiction (and most science fiction), because the authors' willingness to ignore or casually distort the laws of physics made the whole thing seem ridiculous. He found it impossible to suspend disbelief when confronted with such a clearly nonsensical 'fictional universe'.

One of Geoff's many areas of expertise is the physics of clouds, a subject I found especially interesting, since my comics are full of clouds. I'm even working on a graphic novel about a cartographer who attempts to map the sky (complete with its ever-changing landscape of clouds). During one of our meetings we visited Auckland Art Gallery to look at an exhibition of paintings by William Hodges. As we studied the pictures, Geoff assessed the credibility of Hodges's clouds with the merciless eye of a man who knows clouds inside out. It was a sobering experience.

In the end, then, after grappling with Brownian motion, relativity and turbulence, the question I found hardest to answer was the one raised by Picasso. Is art a lie that tells the truth? Or is it just a lie?

So of course that's what I had to write about. Because as any scientist (or writer) knows, it's all about chasing the most difficult questions . . .

P.S. Sorry about the clouds, Geoff . . .

WITI IHIMAERA

Witi Ihimaera is the author of 11 novels and five short-story collections. He belongs to Maori tribes of Gisborne and the East Coast. His first collection of short stories, *Pounamu, Pounamu* (1972), was followed by the novel *Tangi* (1973), the first novel by a Maori writer. Best known internationally as the author of *The Whale Rider* (1987), which was made into a highly acclaimed film (2000), he is a pioneer of world indigenous literature, and works vigorously in other media including non-fiction, film, theatre, opera, art, children's literature and ballet. He is a Professor of English, teaching Maori, Pacific and indigenous texts, and creative writing, at the University of Auckland.

Witi Ihimaera writes:

I was already busy enough, right, when the offer came to join the *Are Angels OK?* project. But flattery can get Bill Manhire anywhere and in a rush of blood to the head I said yes.

Actually, I had carried the idea upon which *Dead of Night* is based in my head for many years, so this was my chance to write it into being. However, five drafts later I knew I was in trouble, and I have to say that the finished story is not the story I envisaged writing. My original idea was to write a story called *The Dinner Party*, in which a group of scientists would have an encounter with an entity in some mathematically defined space. The entity would 'speak' mathematically, and through the encounter would be built up a vocabulary. By the halfway point, the entity would be revealed to be the angel Gabriel, and I had hoped to establish enough of the mathematical vocabulary to begin to substitute the English text with it. By the time the reader reached the two-thirds mark, he or she would have sufficient mathematical vocabulary to be able to read the rest of the story, which, somewhat ambitiously, I had expected to write entirely in mathematics.

Dead of Night will have to suffice. The title comes from a 1940s classic British fantasy film starring Sir Michael Redgrave; this film is also referenced in Simon Singh's book *Big Bang* (2004), one of the many sources consulted in the writing of the story. The narrative of the journey of a modern-day *Endeavour* is modelled on the trip made by Captain Cook in 1769 to observe the transit of Venus in Tahiti; as a New Zealand writer, I have always tried to put New Zealand at the centre of my work, even if, in this case, the narrative involves the universal world of science and mathematics.

The narrative got sidetracked from my original intention because the mathematical switch-over was just too difficult to accomplish. However, I did manage to sketch some of this into *Dead of Night*. That said, the foregrounding of the 'dinner party' conversations on the history of cosmology and, primarily, the cosmological sciences conforms to the original idea: Charles Seife's *Alpha and Omega: The Search for the Beginning and End of the Universe* (2003) was a seminal reference. For better or worse there's always been a strong didactic component in my work, and I had always wanted the story to begin a process of information transmission from the inventory of cosmological history, attaching it to the kinds of New Zealand, Pacific, Maori and indigenous histories that I write about.

In all my career, however, I have never written as many drafts as I have of *Dead of Night*. Nor have I had the experience of the drafts getting progressively *worse* rather than better. By draft 10 I was in sheer terror at the magnitude of my ignorance, vanity and stupidity in having accepted the commission. Around draft 16 I had lost all idea of whether the story and writing were good or not – and the story was still growing like Pinocchio's nose. I kept yelling at it, 'You're supposed to be a short story. Stop growing. Stop. Right. Now.' Readers should count their lucky stars that it isn't any longer.

I have to pay a tribute to my co-writers Howard Carmichael and David Wiltshire for their huge contribution to the story. I'm just not that clever.

I was onto draft five when Howard and I started meeting for coffee or lunch at the University of Auckland. At our first meeting I explained that I wanted the first chapter of Genesis translated into mathematical formulae. I didn't think it would be difficult – after all, the entire Bible has been translated into many languages – but I hadn't realised how sophisticated a job Howard had done until fellow collaborator Rob Ballagh told me; having no mathematical knowledge, I was not able to judge. Howard tried to explain the basis of his translation to me but it is only now, at the end of the writing process, that I have begun to realise how clever – and how much fun – it is. Around draft 12 I finally understood how much of his reputation Howard had invested in the project – after all, I was asking him to buy into a proposition that many physicists don't believe in: the notion that an entity called God created the universe. Throughout the project we continued to meet, and Howard maintained his equanimity while I started losing mine.

Around draft 20, Glenda Lewis from the Royal Society of New Zealand suggested that I should get some cosmologists to look at *Dead of Night*, and I readily agreed. So another team of scientists came to help with the project. One was David Wiltshire, who arrived like space-age cavalry at just the right time to save me from total ignominy. Scientists are busy people, but David set aside so much of his time to ensure the further viability of the story within the scientific context. He was not afraid to involve himself in the text, was generous in his contributions, and added many exciting scenarios and ideas to *Dead of Night* – in particular, the sequence involving *Endeavour*'s arrival at Venus II and the inspiring apotheosis that occurs when the wavelength of the universe leads to its renewal. David also took valuable time from his busy family and lecturing schedule to attach copious explanatory notes to his changes; we were still implementing draft amendments or re-checking the complicated mathematics even at the eleventh hour.

What's great about the collaborations with Howard, David and others is that it has expanded the possibilities of transformation for the story beyond anything I – and I suspect my collaborators, too – could have managed. I don't think they would agree, possibly, with the ending, and there are some things in what they have said that I disagree with; but they have given me permission to take the story where it has gone and I've done the same for them – we have been generous with each other, and for that I thank them. Still, for their sake I will point out that the story should come with the kind of warning that you see on DVDs: 'The material offered by those who have contributed to the story does not necessarily represent their agreement with the philosophical views presented in *Dead of Night*.'

It goes without saying that any validity *Dead of Night* has scientifically is due, entirely, to my collaborators, particularly Howard and David. They must have found me somewhat tiresome, because after every exchange of notes I would redraft the story, only to break yet another law of physics or get the history wrong. 'The rule of the game,' David once emailed me, 'is that whatever you create always has to be reconciled with the known laws of physics.' I had some nice intuitive ideas but that wasn't enough; substantiating or changing them has been in-tuh-resting, to say the least. During the last hectic month of the collaboration, David's interventions were truly inspirational. Emails kept flashing back and forth between us as we tried to meet the deadline, and to align the story with those laws of physics. For instance, at the end of *Endeavour*'s journey I had envisaged the ship arriving at a dramatic wall of fire, but I had to jettison this image because the physics was wrong. Then I wrote that climax as an arrival at a 'cliff', with the universe spilling over it; again, out it went as David patiently explained that the end of the universe is not a place but a time. In the end, David himself wrote the section on the coming of the walls of universes – and a wonderful job he did of it too.

I hope that readers will enjoy the story. Howard emailed after we had finished the final draft, seven days before deadline, to say, 'Now we wait for accolades or rotten tomatoes.' Frankly, I never expect accolades and, if there are rotten tomatoes, I think Howard was probably referring to the way in which I have viewed the history of the cosmological sciences, some of the scientific and non-scientific challenges and claims I have made, and views I have expressed: bring them on. In Maoridom we appreciate korero.

What I am relieved about is that the story in most of its aspects has some validity for both the scientific and the literary audiences that we serve. I like to think of it as an embedded text, having ideas, inflections, references, theories and propositions that we've all contributed. Some of the embedded material I am only beginning to understand – like the Wheeler–DeWitt equation of quantum cosmology – but I am pleased that my colleagues have ensured an informed text for their constituencies. I also hope my non-scientific colleagues will consider the viability of the story. During its writing, I was invited to a dinner involving senior New Zealand clerics to honour the visiting Anglican Archbishop of South Africa, Njongonkulu Ndungane. People have a habit of asking me what I am doing, and I told them about *Dead of Night* and the proposition that the language of God is mathematics. It was as if a dead white rabbit had been thrown onto the table, and the subject was quickly changed.

Here's what the collaboration was like – being in a tandem parachute jump. Usually, as a writer, I'd be the one pulling on the parachute's cords and choosing where to land. But in this collaboration with Howard, David and others, I was often the tandem passenger, one to be humoured and forgiven for his appalling lack of knowledge – and you wouldn't believe how difficult it was to swap places. However, I realised I had to trust my collaborators, even when I could barely understand their esoteric science-speak. I kept on saying to myself through gritted teeth, 'You can do this. So we're going in now? Ookayyy . . .' Letting them take me in was, in the end, tremendously exhilarating, and we have landed in so many squares where landings were not originally planned.

I have been honoured to have had the opportunity to learn a little more about the world of science, in particular physics and cosmology. As a creative writer, I have in fact explored scientific themes in earlier work, most notably in an opera entitled *Galileo* which I wrote for composer John Rimmer, and which premièred in New Zealand in 2002. However, I still know absolutely nothing about science. I agonised over the ending, particularly the appearance of the double helix – I like bold gestures, but wasn't too sure about making this one. I asked Jane Parkin for her opinion and she mentioned that, coincidentally, newspapers had reported the discovery of a cosmic nebula twisted like the double helix of DNA. I checked, and sure enough there it was, so I decided to keep the double nebula in and maintain the 'open' ending. I am grateful, however, that the editors have allowed me to publish the original 'closed' or 'circular' ending for those who are interested.

In terms of the imaginative possibilities that have been opened up, I'm not daunted by the complex intellectual territory. Like Captain Walter Craig in my story, if I ever had another shot at life, I would want to be a scientist. I pay tribute to the scientific community in New Zealand for their great work.

My debt to Howard and David I hope is clear. But others such as Rob Ballagh, Professor, Physics Department, University of Otago; Matt Visser, Professor, School of Mathematics, Statistics and Computer Science, Victoria University of Wellington; and Michael Walker, Associate Professor, School of Biological Sciences, University of Auckland, have generously read drafts of *Dead of Night* and provided valuable comments and suggestions. Stephen Pritchard, Unix Systems Engineer, Hewlett Packard, Wellington, was helpful with information on speed of light travel. Jane Parkin edited the text. Any errors of fact or imagination, however, are my own.

The Original Ending of *Dead of Night*

The mathematical formulae stop.

Mrs Cortland presses Miss O'Hara's hands reassuringly. 'We'll be all right, my dear.'

The *Endeavour* rocks. The three aunties are like lifeboats beside it. They are chattering as if they know what is going on.

'*Up, up and away, my beautiful, my beautiful balloon!*' Aunti-3 sings. Her voice is ecstatic. 'Yes,' Aunti-1 and Aunti-2 respond. 'We've done our job.'

'Captain,' Hemi says, 'I'm getting readings of huge energy forces coalescing all around us. Something is happening.'

The ship rocks, and rocks again. Then everything is happening at once. Echoes of the whole titanic history of the universe are crowding in, pushing and jostling with each other, fighting for the last sliver of space. Every movement ever made. Every word ever spoken. Every television show ever broadcast. Every ray of starlight ever shone. Space is collapsing under its own weight.

'It won't be long before the gravitational waves overwhelm us,' Captain Craig says. In his mind's eye he sees a trillion black holes in collision. Tears spring to his eyes at the thought of all that human history – whatever the self-destructiveness, there had also been so much *life,* so much hope, so many dreams achieved and triumphs witnessed. As one of the last survivors, Captain Craig raises his voice in poroporoaki, in defiant tribute to the generations upon generations of men and women who had lived in this life and world:

Tena koutou nga iwi katoa o te Ao,
Te Huinga o te Kahurangi,
Tena koutou –

Hemi's voice comes, soothing, to Captain Craig. 'Ten seconds to the first impact from the universe's shock waves. Nine, eight, seven . . . Sir, time to open the codicil to your secret instructions.'

'Open,' Captain Craig orders. On all screens there appears one word:

RESET.

'Four, three, two, one –'
    'Do as ordered,' Captain Craig says.
    'Zero.'

10.

Time fades to nothing.
    Real nothing. Not even space. Outside? Outside there is no outside.
    Space is time and time is space. Space enough, perhaps?
Then:

$H |\Psi> = 0$

The ship is like a bloodied jawbone thrown through the air. The energy is more than 10 billion billion billion nuclear detonations. It roars over the *Endeavour*. In a trice, the ship is incinerated. Her avatar, Hemi, gives a huge, deep sigh. The glorious aunties are like angels on fire, fluttering into oblivion.

Captain Craig feels an intense pain. He looks at Mrs Cortland, Miss O'Hara, Monsignor Frère, Professor Van Straaten and Dr Foley, wanting to reassure them. But of what? Just before the endorphins kick in, he has a regretful thought:

'But we didn't have time to say goodbye to each other.'

And he is falling.

He feels a dizzying rush of acceleration, as if he is being sucked at headlong speed down a tunnel of dazzling light. Onward and onward he roars, and the sensation is so delirious that he wants to laugh and laugh.

All of a sudden he is through the tunnel and suspended above the blackness, watching the primordial fireball and the way its wave of light is moving so quickly away from him. He becomes frightened and closes his eyes.

And he is falling.

When he opens his eyes he sees a wild landscape in the country of his birth, New Zealand. It is so familiar to him that he laughs with relief. Suddenly a mist descends and he is lost in it.

This has happened before, he thinks. The mist opens and he sees three elderly women from the village walking in front of him. He runs after them. One of them turns and asks, 'Went the day well, sir?' The second says, 'Not you again.' And the third says, 'You're always losing direction. Well, you're almost at your destination, lad. There it is.' She points to a faraway farmhouse on the other side of the valley. 'You had better make haste,' the old women say. 'Night is coming and, with it, a fierce storm.'

He walks to the farmhouse. A light is coming from the window and, framed within the light, someone is watching him. As he approaches, the door opens. Mrs Cortland is there.

'Thank goodness you were able to make it home before dark,' she says. 'Come in, come in.'

He enters the house and sees that Monsignor Frère, Dr Foley,

Miss O'Hara and Professor Van Straaten are having a cup of tea and chatting. 'Let the Captain have a place by the fire,' Mrs Cortland scolds. 'He'll catch his death. There's room for one more.'

The five people in the room smile at Captain Craig as if they have known him all their lives. Monsignor Frère comes to join him. 'Don't be afraid,' he says. 'There are some questions that science cannot answer. They may know what, how and when, but while faith and reason have co-existed in scientists as notable as Isaac Newton and Albert Einstein, only theologists know *why*.'

The Captain shifts on his feet nervously. Mrs Cortland brings him a cup of hot tea. 'This will bring you back to life,' she says.

A few moments later, Professor Van Straaten comes to talk to him. 'You shouldn't believe everything you see,' he says. 'At the end of the universe things still go bump in the night and every grave, opening wide, lets forth its sprites and demons. But Fate is always kind. It gives to those who love their perfect longing.'

'Do I know you?' the Captain asks the Professor. 'Do I know any of you?'

'Of course you do!' Professor Van Straaten laughs. But he is uncertain, and his laughter fades away into bewilderment. 'Because if you don't, then who are we?'

Monsignor Frère winks at him. 'Interesting, isn't it?'

Mrs Cortland claps her hands for attention. 'It's time to go,' she says. 'Dr Foley and Miss O'Hara, are you ready? The next great adventure is about to begin. As for the Captain, he has his own journey. Goodbye, my dear.'

And he is falling.

Suddenly Captain Craig finds himself in a small white room. He is alone. He is naked. How did he get here? Who is he? Why is he here? He begins to scream and pound on the walls. He falls to the floor in a foetal position, howling, hugging himself and weeping.

Time is limitless. How long has he been here? He does not know. He sleeps, and 15 billion light-years go by. For one brief second he feels butterflies brushing his eyelids, and knows his wife and children have flitted by in his dreams. More light-years go by.

Finally, he awakes. He sees that there is a wardrobe in the room. In the wardrobe is a high-necked suit. With a sigh he dresses in the suit.

A door appears in the room. Above the door is a clock. The time is just before midnight.

Captain Craig finishes dressing. He sees a mirror on the wall. He inspects his appearance. Combs his hair. Smooths out his trousers.

Takes a deep breath. Walks to the door.

Turns the doorknob.

Opens the door.

The table is set for six.

It is circular. All the guests when seated will be equidistant from each other. The host, Captain Walter Craig, has invited them for pre-dinner drinks at 7.30. Time, of course, has long lost all real meaning, but on board the ship a 24-hour day is still observed; it continues to locate, structure and define, and, by setting a beginning for the journey, has enabled the Captain and crew to calculate the various coordinates they have reached as the journey has progressed.

The ship is an Artificial Intelligence called the *Endeavour*. Powered by its avatar, Hemi, it is a silent celestial angel in solitary flight through a sea of stars, cleaving through the blackness, serene and powerful, its light-wings at full extension, across the dead of night.

LLOYD JONES

Lloyd Jones grew up in the Hutt Valley and is perhaps best known for his novel *The Book of Fame*, which both records and imagines the 1905 All Black tour of Europe. (It won the Deutz Medal for Fiction at the 2001 Montana New Zealand Book Awards, then in 2003 won the Tasmania Pacific Fiction Prize.) Lloyd is the author of many other novels and collections of stories, and has recently written and published work for children, in particular the best-selling *Everything You Need to Know about the World by Simon Eliot*. He is also the publisher of an important series of essays by New Zealand writers through his Four Winds Press.

Lloyd Jones writes:

I'm struck by how ghetto-ised writers inevitably risk becoming and how incredibly helpful this project was in wresting me (especially) out of the worn cracks. I wasn't the most brilliant student at high school (I was in a class of ratbags for most of it). Yet, for all that, even at its most speculative I find it quite easy to talk about physics with physicists, because in the end we reach for the same language of metaphor.

My initial inquiry was into the difference between 'space' and 'nothing'. This quickly and inevitably led to discussions of, well, just about everything to do with 'out there' – the universe. I didn't feel obliged to stick to the trail. I went wherever my interest led me. My reading followed suit, and Jeff Tallon and Phil Butler were willing and engaging colleagues.

This scatter-gun approach encouraged a sort of global view and inevitably led me on to the idea of entanglement of space and time. I filled notebooks with the things I was learning. I also kept a notebook that charted my own responses (from outright dumbness to moments of fleeting insight). This was the fun and easy part. The challenge was to find the hook for the writing part. I took a wide and liberal view of 'imaginative response'. I knew what I didn't want to do, and that was merely to illustrate. In all of this the lightning rod for me was first hearing of 'elsewhen' in a lecture by Matt Visser. My response was pretty much as revealed in my piece. You will guess by my list of 'discoverers' that my reading was broad. I read the science, but at the same time I was skirting around for an imaginative way to respond. I felt I hit gold when I discovered what Gödel had to say about the value of 'fables'. The real technical challenge was taking to heart his observation of the absence of present/past/future considerations in the cosmos. The space in between the stars isn't rushing to catch a train, never has, never will. I began to play around with stories that would reject the usual ordering of time.

This was the truly exciting part for me, to be able to be wilful and even 'irresponsible' with regard to the reader's comfort. And of course, where the idea of entanglement comes in will be obvious. The tales intersect, refer back to one another (in some instances). There is an echo going on. Though this will be more obvious when I finish in another two years' time. This project is the start of a larger and more ambitious one. I had no idea what I was getting myself into, but I'm sure glad I did. The project continues.

ELIZABETH KNOX

Elizabeth Knox is best known for her acclaimed novel *The Vintner's Luck*, which has sold over 100,000 copies worldwide and won the Deutz Medal at the 1999 Montana New Zealand Book Awards. Her other novels are *After Z-Hour, Treasure, Glamour and the Sea, Black Oxen, Billie's Kiss* and *Daylight. The High Jump: A New Zealand Childhood* gathers together the three novellas *Paremata, Pomare* and *Tawa*. Recently *Dreamhunter*, the first of a two-book series for young adults, has been published internationally. She lives in Wellington with her husband and son, and has just finished *Dreamquake*, the concluding part of her *Dreamhunter Duet*.

Elizabeth Knox writes:

When Bill Manhire sent me a list of topics for *Are Angels OK?* I was very quick in picking time travel. I was thinking about time travel, writing my *Dreamhunter Duet*, and I've liked time travel stories since at 13 I read Clifford D. Simak's *Time and Again*. In *Time and Again* a man called Asher Sutton finds a copy of a book that he has apparently written, a book that, he discovers, is the founding document and sacred text of a future civilisation. The future Sutton's book has engendered isn't very rosy for humans (but is for androids). Sutton is appalled. He tries to resist being the person he is and thinking the thoughts he has so that he'll never write the book. Humans from the future try to kill him. Androids from the future protect him – with great tenderness. Ultimately Sutton becomes resigned to his fate and, as a serene and philosophical old man, sits down and writes his book.

Like Simak's *Time and Again*, my favourite time travel stories deal with a universe-bending desire to fix things, with the impossibility of fixing things and, by inference, with redemption, salvation and destiny. Most of these favourites involve time travel to the past within the time traveller's own lifetime.

I started to do some reading for the project. I didn't read about time travel in fiction. My head was already full of examples from films and books, the stories that physics writers use to explain how time travel might work. How it might work isn't about how to do it – that's a separate matter. All these stories must re-imagine causality. They pose questions like: are we living in a universe where what happens is what was always going to happen? (*12 Monkeys, Bill and Ted's Excellent Adventure.*) Are we living in a universe we can pollute with paradoxical repetition? (The 2004 film *Primer* is the very best example I've seen of this; it is fiendishly, feverishly, nauseatingly complex – and I love it!) Or are we living in a universe where each alteration of events in a timeline generates another timeline?

There are far fewer stories providing examples of the last. It is one thing to imagine parallel worlds, like those in the Jet Li film *The One*, or in Diana Wynne Jones's Chrestomanci books. But where there are many worlds because new timelines are generated with each change – well – the opportunities for satisfying stories are limited. In these stories the right hand won't know what the left is doing, or has done, or that it is right because of something the left has done wrong. For a many-worlds time travel story to mean much the possibility of a self-consistent universe would have to be built into it first, so that the resolution of the story would rely on the defusing of the tragic inevitability of self-consistency by the knowledge that things had happened differently, even if they happened in timeline B rather than timeline A.

All this was what I was thinking about. Then, mid-year, I luckily met Gerry Gilmore, the deputy director of the Institute of Astronomy in Cambridge. I told him that my subject for *Are Angels OK?* was time travel and, yes, I knew about the speed of light, space being curved, and 'the twins effect' (the stay-at-home ages 20 years to the three years of the twin travelling at close to the speed of light). I pretty much grasped all that stuff, I said, though don't get me onto flashlights shone towards the front of ships travelling at close to the speed of light. However, I said to Gerry Gilmore, what I'm interested in is time travel to the past. I liked those stories where people try to fix things. I wanted to find something to say about why we like those stories.

I said that I wasn't really interested in 'time-travelling tourists in history'. Though I did love Ray Bradbury's story about the time travellers who are told to shout 'Free Barabbas!' in order not to change the course of history, until one man realises that much of the crowd is comprised of tourists like himself, religiously following the tour company's instructions. (The events of this Bradbury story are, of course, one argument for how we can know that no one in our future will invent a return-trip method of time travel – if they are going to, then the great moments of history would be crowded with visitors from all the periods post the invention, all of them probably behaving themselves, peeing into funnels like Antarctic tourists, and shouting 'Free Barabbas!' on cue.)

I told Gerry Gilmore that I planned to write an essay. 'Because I don't write short fiction.' Then we talked briefly about Tachyons, the theoretical quantum particles that travel backwards in time. And then Gerry told me that time travel to the past was, he thought, impossible. First you would have to have a black hole, then a wormhole created by that black hole, then something to keep the tunnel of the wormhole open. And, even then, given the ravening gravity, how could anyone get anywhere near it? Then he said that, of course, if the first two conditions were met – black hole to wormhole – then it might be possible to send *information* back to the past. Not a traveller, but a calling card.

Earlier that evening, after Gerry's talk, when I was standing around with the Royal Society people and Vic physicists waiting to go out to dinner, I asked Paul Callaghan what kind of people buttonhole physicists after their lectures. Paul told me that most of them are shy and only want some point clarified, and some want to say hello from Auntie Gladys. 'I'm sure that happens to writers too,' Paul said. Then he said, 'But there are the ones who have *theories* and say that they just need someone to do the maths.'

It was these two things – the idea of information sent back to the past, and the image of a sweaty-palmed person with 'a theory' and a ravening need for attention not from people so much as from external reality itself – it was these two things that began to come together for me.

Not that I knew it. I continued to mutter to Bill that I didn't know what to write about. Then, some months later, I was asked to sit on a panel – three physicists, three writers, chaired by Kim Hill – and talk about what I was doing for the project. Well – what I was doing was writing *Dreamquake*, which is, among other things, a time travel story. But I still had no conscious apprehension of what I intended to do for *Are Angels OK?*

What we come to write is almost always due to internal differences, tectonic faults in our personalities that are there already – always there – and owe next-to-nothing to our apprehended intentions.

In preparation for the panel I read Paul Davies's *How To Build a Time Machine*, a book I bought a few Christmases back for one of my nieces (who is now studying physics at Vic, much to my delight). I read Richard Gott's excellent *Time Travel in Einstein's Universe*. I was able, when Kim asked me to explain wormholes, to give a 'then and then and then' explanation, and to congratulate myself quietly on not giving the zen answer, which believe me I felt like giving after cramming the Gott: 'What is not a wormhole, grasshopper?'

I sat on the panel, and talked to the physicists afterwards – though I was still harping fruitlessly on why time travel was interesting to me. Then I went home and forgot all about it again.

Then I ran out of time.

And because I ran out of time I could only write about the thing I always avoid writing about, glancing against it only occasionally, though astute readers – like Karl Stead in his 'Knox's Oxen' – do sometimes notice that it is the thing at the heart of all my other things. I have to say that rather than write about it it seemed easier to disappoint Bill and pull out of the project altogether, and, believe me, disappointing Bill isn't often the easier option. I thought there must be something *else* I could do, a story I could tell, a different approach – yes, I had sunk so low as to be thinking in terms of approaches, as though it were a task like landing a plane, and I was working *with* gravity instead of consenting to fall.

In the end I didn't swerve, though there may be people who say that science fiction is swerving and the only way to tell certain stories properly is to do it in realist fiction or in memoir, but I say to those people that they should read more physics.

My story tackles the question that had been regarding me unblinkingly for months, waiting for me to notice that it was my question, not 'ours', and came out of my life, not culture. Why am

I so interested in certain kinds of time travel stories? Because there are things I wish I'd known sooner. So, somewhere between talking to Gerry Gilmore and sitting on the panel, while trying to work out how I could get away with using my physics only metaphorically – that is, how to avoid anything so distasteful as writing science fiction – I had scribbled this on one of the end papers of the Davies book: 'Quantum tunnelling. The foam of space-time imagined as the spongy matter of the younger brother's brain. There might be a wormhole, an infinitessimal virtual wormhole through which his present understanding could make its way back in time to tell the boy he was to *expect nothing and hope for nothing*, to stop casting the line of his own gaze where it won't be caught, to stop hailing for help or anything else till all he has learned is how to find a lifetime's worth of heartless people to vainly persist in loving for his lifetime. Surely what the man knows now isn't so *big* that it couldn't find its way back to the boy he was?'

Quantum tunnelling, cosmic strings, wormholes whose gravity is too great to let them anywhere near our solar system – I considered all these things. Then I called Paul Callaghan and said, 'I need some *thing* that will send information back in time. Can you help?' I went to see him and we spent an entertaining hour, me taking notes and he bouncing about and drawing diagrams on his whiteboard. I had come to Paul with a time machine design I'd read about. The Tippler machine is the superdense cylinder spinning on its axis at half the speed of light. A Tippler machine could create a zone of null time, and one of backwards-running time, because its gravity would bend space-time. That was Frank Tippler's theory. The Tippler machine is centuries beyond our current technology, since it is kilometres long and must be constructed in space. The story I was writing had to be, at its latest point, only a few years on from now. 'My Tippler machine is going to have to have just turned up,' I told Paul. 'As a piece of drifting alien technology.'

Paul pulled a face. 'Do you want there to be aliens?'

'No. Aliens are a bridge too far. But can I avoid them?'

(It is quite funny that, for Paul, angels are OK, but aliens irritate.)

Between us we designed the thing that appears in my story as The Deity, aka Mr Ed. We didn't get it quite right, of course, and several days later I emailed Richard Easther, a New Zealand cosmologist in New York, whose partner Jolisa Gracewood had said that if I got stuck maybe he could help. I sent Richard a description of my wormhole germ and, later, one question: 'Do I mean anti-gravity, not antimatter?' To which Richard responded that, yes, I meant anti-gravity, and, 'The standard geeky joke about this sort of thing is that while one can describe such a situation theoretically, it would have to be constructed from "unobtainium", which rules out its existence in practice.'

Richard had provided a title for my story: 'Unobtainium', the stuff you can't find anywhere.

Here it is. Like Simak's hero I gave in to being the person I am and thinking the thoughts I have and wrote perhaps not what I was always going to write but what I am always writing.

MARGARET MAHY

Margaret Mahy is perhaps New Zealand's best known and most loved writer. Her many picture books for smaller children (the first, *A Lion in the Meadow*, was published in 1969) keep company with remarkable works of fiction for junior and young adult readers. Her books have been published and translated around the world, and she is the recipient of many awards, including the Carnegie Medal (twice), the Prime Minister's Award for Literary Achievement, and – most recently – the Hans Christian Andersen Award, while her physics-based novel *The Catalogue of the Universe* received the 2005 Canadian Children's Literature Association Phoenix Award. Margaret Mahy is a member of the Order of New Zealand. Her book of essays, *A Dissolving Ghost*, is published by Victoria University Press.

Margaret Mahy writes:

Two or three years ago I had a calendar on my wall – an Einstein calendar, actually – each month marked by a photograph and a worthy quote from this remarkable man. The one picture I have saved – I can see it now from just beyond my computer – has Albert looking thoughtful but relaxed as he sits in something like a deck chair and the words below the picture quote him as saying, 'When I consider myself and my methods of thought, I come to the conclusion that the gift of fantasy meant more to me than my talent for absorbing positive knowledge.'

For some reason this does not seem to be quite what we expect from a great physicist, since physics, and science generally, seem established to hold fantasy at bay. Nevertheless, perhaps one of fantasy's functions is to pinpoint and define areas in our understanding that might be initially seen as impossibility but which give liberty to human speculation. Certainly, this quote seems to acknowledge speculative imagination as source not only of art but of science too, even if theories thrown up by creative guesswork must then be subject to tests, must undergo intricate examinations which enable forecasts to be made, and subsequently fulfilled or nullified. The word 'accuracy' is often seen as a clinical one, but it is the accuracy of certain images, certain words, certain facts that makes them moving. Accuracy is not just a chilly aspect of learning but can be an aspect of beauty. And in the same way it is, apparently, the imaginative fantasy of certain scientific statements which somehow projects the thinker into new, and possibly provable, speculations. As the speculation is tested and moves towards becoming what we describe as a truth, it also becomes something from which we derive fulfilment – something to which we may legitimately have something like a passionate, even a poetic, response.

I was below average at science at school and considerably below average when it came to mathematics, that dancing partner of science. However, I did have, along with many of my fellow pupils, a distinct curiosity about the way the universe was put together, and just how its various systems, small or grand, might actually work. That interest in astronomy tended to dominate for many years. I can well remember buying a telescope kitset (an object lens and three eyepieces) from a local camera shop in Whakatane. My uncle next door (a plumber) made me a tin tube of the designated length. My father helped me put stops in place along the tube and in due course I looked through it at the moon. Almost the first thing I saw as I adjusted the focus was the crater Copernicus. Later, looking at Jupiter, I was able to make out the four Galilean moons, and though I did not know, straight off, which one was Io, which Callipso, which was Ganymede or Europa, I did at least know their names, and I will never forget the emotion that flooded me when I looked up out of my backyard in Whakatane and actually saw them. I had believed what I had been told in books, that they were actually there and had played a part in establishing an extended understanding of the immediate solar system, but now their existence was confirmed in a personal way. Excitement was part of my response . . . but it was more than excitement. There was the feeling that I had moved into a new relationship with that universe. Over following nights I watched those moons change place, sometimes eclipsing each other in their dance around this huge planet. I also located Saturn and made out its rings. Later I saw Mars as a reddish pinhead . . . rather negligible through my telescope, except that I knew I was looking at the closest planet to the Earth, a planet that appeared, dramatically, on the cover of many science fiction magazines. And I watched Venus move from being a tiny, bright crescent to being a full sphere then back to crescent again.

For someone who is not a scientist, the effort to cross over and to understand something of what science – of what physics – has to tell is always complicated by doubt. Yet isn't an approximate understanding better than no understanding at all? The problem for someone like me is how to tell just how approximate the approximate truth you are reading might happen to be. And what happens when one reaches the limit of imagination? I have always unconsciously assumed, until recently, that imagination has no limits, and that, in its speculation, it can jump away from any truth in a wide variety of directions without detaching itself from that central anchor of truth. But some of the things that have been said to me recently make me think that imagination does have limits, and that, in some ways, we are approaching those limits.

For example, Andrew Wilson – the physicist who was inflicted with me as a partner – and I found ourselves talking, perhaps inevitably, about the Big Bang; and he said, quite casually, that the Big Bang had created space, and I have gone on from that to try to understand just what space might be. Simple absence – the absence of everything. It may have an identity and yet it is simultaneously nothing . . . there is a curious contradiction here.

There is a human dilemma involved. I want to look at things and, without necessarily pulling all the facts that distinguish them into my mind, to feel the underlying power and poetry of their factual nature, the accuracy of their inherent poetry, the poetry of their accuracy. Such scientific knowledge as I have managed to achieve enriches my view of the world; but I still have friends and acquaintances who see these ideas as somehow reducing poetic truth.

I own a book called *Einstein's Heroes* by Robyn Arianrhod, which I look at from time to time. I paid good money for it and therefore feel the information it contains is mine, but I haven't actually read it yet!

'Imagine,' says the blurb on the back, 'you are fluent in a language so powerful that when you write it down it mysteriously takes on a life of its own beyond your thoughts or control. A language of prophecy with which you can accurately describe things you cannot yet see or even imagine.' I am intimidated by the thought of reading a book about mathematics, though I do plan to read it if ever I have time and solitude; and I imagine myself reading it very slowly, holding a finger under every line I am reading (as small children do) so that the words don't take on lives of their own and squirm away from me. And once again, at the same time, reading that blurb, I thought the description of the language the book deals with might apply to any language, for all words can take on a life beyond thoughts and control. They may be constrained by everyday necessity to a considerable extent, but flowing from the end of the poet's ballpoint pen they too continually become instruments of revelation, and can describe things one cannot yet see or even imagine – until one has read them, that is.

Another book I have – also rather out of date by now, but still useful – is called *The Catalogue of the Universe*. Over the years I dipped into this book, read pieces of it, and in due course gave a young adult novel I wrote the identical name. Its hero is called Tycho, and of all the characters I have invented I often feel he has the most in common with me. The disclosures of science seem to him simultaneously clarifications but also vehicles that carry him deeper and still deeper into mystery. And some of the experiences and responses described in the book are certainly my own.

'The thing is common sense and truth don't match – not all the time.'

'That's absolutely like you – to say a thing like that!' Angela said vigorously. 'Come back to the car and tell me as we go.'

'Common sense is tidy and truth's untidy,' Tycho declared, following her obediently, but though he sounded certain of what he was saying he was actually working something out. He was really talking to himself. 'Common sense is very neat and it's easy to see, and it's sort of symmetrical, bowling along in the open and looking completely real, but it's only common sense – whereas actual truth wobbles and hides.'

'Wobbles and hides, wobbles and hides!' chanted Angela, smiling back at him. 'You've invented a chorus. Wobbling, hiding truth.'

'Truth's furtive,' Tycho shouted up to her, beginning to enjoy his ideas, all the more because he believed them, and putting them into words gave him power over them.

'It's an ellipse not a circle,' Angela cried back, catching his ideas, inspired by a story Tycho had once told her about Kepler working for years and years to define the orbit of Mars because he believed the orbit of a planet must be a circle. It had seemed so logical for it to be a circle with the sun at its centre, but Mars had tricked him by moving in a slight ellipse. 'It's got two focuses,' Angela said, remembering.

Tycho was thrilled with this idea.

'Spot on!' he said, stopping in his tracks. 'The world's left handed. Planets move in ellipses, parity isn't preserved and the square root of two is an irrational number.'

I do think human beings have an implicit response to symmetry, and I understand that there is a great deal of approximate balance and symmetry in the physical world which can give pleasure and which deserves recognition. At the same time, I think there is something liberating about the idea that symmetry is not universally observed. The thought of the apparent irregularity of the cobalt nucleus gives me pleasure too, for I somehow like to think that, though I am as anxious to contain the world in a cage of prediction and understanding, I also enjoy the thought that existence does, in certain sly ways, escape from that particular cage no matter how beautiful and secure it might happen to be . . .

Editors do not always welcome the intrusion of imaginative scientific speculation in children's stories, but I always feel a particular affection for *The Catalogue of the Universe*, that novel with a stolen title. Looking through the book again after many years, I came on a piece which describes Tycho setting up his home-made telescope, and I realise it is a description of my own experiences with my telescope. The only creative invention about it is the way I have passed my experiences on to a character in the book.

And in the end, the heroine, Angela, makes the hero, who is slightly shorter than she is, stand on the book ... stand on *The Catalogue of the Universe* ... so that they can kiss as equals. 'Of all the people in the world,' says Angela, 'you're the one who stands on *The Catalogue of the Universe* every minute of the day.'

I mention all this to emphasise something that everyone knows already: that art (painting, writing, carving, composing) and science (cosmology, physics or whatever) are not closed-off compounds, but in their various ways are part of the human flow of conjecture. Each discipline has its own necessities which must be respected, but there are legitimate conjunctions too. Though I am, and always will be, excluded from the truest knowledge of physics, I am obviously not above joking around its edges, finding some sort of admittance in acknowledging what I know but simultaneously what I don't know and will never know. I am not above building what I do understand of it into myself (always bearing in mind firstly the approximations of my understanding and the responsibility I have to improve that understanding as far as possible ... to be as accurate as I can, in fact, and to use this accuracy to move as deeply as is possible for me into its mysteries and images and challenges). I do this in a way that I think is both personal and commonplace – for, after all, in a way we are all, to a considerable extent, what we are able to find out about ourselves, and that involves an understanding of what we are and where we are ... assemblies of subatomic particles in an expanding universe ... assemblies capable of entertaining happiness and anguish, curiosity and wonder, all of which mark our advance into the world and which we build back into ourselves ... assemblies that respond to the world and its underlying enigmas with the variable energy we describe as art.

So! Let's hook into the flow! Let's think and find out. And let's dance.

BILL MANHIRE

Bill Manhire was born in Invercargill, New Zealand, did postgraduate work on the Old Norse sagas, and now directs the International Institute of Modern Letters which manages Victoria University's well-known creative writing programme. His many books of poetry include, most recently, *Lifted*, and a *Collected Poems*, both published by Victoria University Press. He has edited a groundbreaking anthology of imaginative writing about Antarctica, *The Wide White Page*, and the poetry anthology *121 New Zealand Poems*. He was the inaugural Te Mata Estate New Zealand Poet Laureate.

VINCENT O'SULLIVAN

The author of many volumes of poetry, Vincent O'Sullivan is also well known for his work as a novelist (*Let the River Stand* and *Believers to the Bright Coast*) and short story writer (most recently *Pictures by Goya*), and for his writing for theatre. He has written a biography of John Mulgan and is a distinguished Katherine Mansfield scholar. He is the editor of a number of major anthologies of New Zealand writing. For many years he was Professor of English at Victoria University, where he now holds an *emeritus* chair. In 2000 he was made a Distinguished Companion of the New Zealand Order of Merit in recognition of his services to literature.

Vincent O'Sullivan writes:

What sparked my first interest in this project was hearing Paul Callaghan give a talk at the Stout Centre, in which he explained with remarkable clarity, and in a way the lay person readily could grasp, the kind of nanophysics he worked in. It was hearing that science-for-the-non-scientific lecture, and my later meetings with Paul, that brought home to me how the 'small places' of physics were as fascinating as the cosmic depths. I wondered if it was possible to write poems that responded to that strangely compelling area of science, and to physics generally. Poems, I mean, that stood up on their own terms, and were not simply transcripts of 'science-speak'. I was intrigued by what would happen if poetry tried to take on board at least something from that (to me) unfamiliar world.

I realised at once that this was not remotely to engage in any scientific way with things, but simply to look from an uninformed perspective at a vast and often incomprehensible method of discovery. All I could hope for was to fire a few of its pellets, as it were, into those areas poetry is usually concerned with – the associative flow of language, verbal and rhythmic patterns, games of analogy, and what Wallace Stevens calls 'the intricate evasions of as'. Quite simply, I asked myself if finding out a little about physics could be a stimulus to poetry, and whether this would bring something to poetry that wouldn't otherwise be there?

There were a couple of things I knew I *didn't* want to do. I had no interest in the kind of sci-fi writing that takes a scientific postulate and carries it through into fantasising situations, where the science merely becomes a narrative device. Nor did I want just to find metaphoric substitutes for what interests science – drawing pictures, in other words, saying, 'Let's imagine what a pion looks like'. It was pretty apparent that all I might do, really, was to try to show what happens when a poet is presented with scientific ideas or experiences, and attempts to fit them, somehow, to his own way of responding to things – what is it that fascinates me, how can I say something about that fascination, in the forms a poet usually employs?

My kick-start was the kind of anecdote that histories of science often include – Lavoisier sent to the guillotine by a jealous, less gifted scientist, the Curies defining radiation even as it was destroying them. Stories that carried some of the excitement and intellectual passion of science on a human, anecdotal level. That led to poems that considered the strangeness, from a mere writer's point of view, of the worlds physicists engage with. To hint at such things I took the biographical detail of Newton sometimes sitting on his bed for hours, unable to get dressed, as he followed through a line of thought. The fact that nature is controlled by knowledge, that the mind in a sense creates the nature it reveals, were notions that interested me – not science by any means, you might say, but things I would not have thought of trying to write about had I not been reading about science.

The first group of poems is really no more than that, poems that have a shot at responding to personalities and contexts that in a general way declare: 'Thinking about what scientists do, even to the limited extent I understand, makes me want to say something like this.' 'Into the depths . . .' begins to do a little more, with the notion that art and science may come at nature and experience from very different approaches, but with a shared intention to define and clarify. The references to Dante and the Southern Cross are from the *Purgatorio*, Canto I, and to the Eternal Rose, from the *Paradiso*, Canto XXX. The famous statue of Poseidon in the National Museum in Athens, with its bronze arms impressively spread wide, becomes the image for what covers both art and science, the drive to discover reality through form.

'A Simple Man's Quartet', rather tongue-in-cheek, takes an actual physics experiment and presents it in a group of three poems, and a fourth that is more reflective on science in general, really just to see if this might be done. In order for it not to take itself too seriously, rap is used for the first poem; the discrepancy between the form and the content is obviously a bit of a joke. There is also enough irony, I hope, in the third piece to underline that this is obviously a *game*, as all poetry is to some extent. Science of course doesn't need this sort of thing, but it's interesting poetically to give it a go. The long note with that set of poems is necessary to give the verse its context.

The group called 'w.w.w.' relate, one way or another, to the world wide web of spiders, from the Arachne myth in Ovid to details of their physiology, the micro-processes, and how we respond to what we know of them.

CHRIS PRICE

Chris Price has worked in publishing, edited the literary magazine *Landfall* for much of the 1990s, and was for many years coordinator of Writers and Readers Week for the New Zealand International Arts Festival. She currently teaches the poetry workshop at the International Institute of Modern Letters, Victoria University of Wellington. Her collection *Husk* (Auckland University Press), which includes a section of poems on science-related themes, won the prize for Best First Book of Poetry in the Montana New Zealand Book Awards 2003. Her second book, a collage of fiction, essay and biography called *Brief Lives,* will be published in 2006.

Chris Price writes:

Most people have doctors and dentists; I have taken great delight in being able to say, 'I'm going to see my physicist.' I began with three physicists in two different locations, and the general brief to tackle 'the physics of life'. Early visits to physics labs at Victoria and Waikato universities were soon followed by a lot of reading – physics, neuroscience, biography, poetry, *New Scientist* and so on. I also attended the Royal Society's excellent VIP Physics Class, better known (to me at least) as 'Physics for Dummies', which had the salutary effect of making me feel I'd been going about the world with only half my brain switched on. I fell in love with Einstein (a habit of mine, these unrequited relationships with difficult-to-love men). At the beginning, I imagined I would write something that touched directly on the work of the local physicists (and that may yet happen). But I soon began to feel that, rather than concentrating on a specific field of physics, I wanted to make a piece of work in which the conversation between 'the two cultures' was part of the deep structure.

Aside from the difficulties of wrangling a vast and diverse collection of information into some kind of coherence and shape (and without the project deadline this piece might have been a great deal longer), the greatest challenge was staying close to the physics behind the subjects that interested me. While I attended lectures on 'The Electrical Properties of Neurons', for example, it was what happened further up the line as a consequence of those electrical properties that came to preoccupy me most. Early on, I found myself writing poems about odd neurological disorders such as Capgras's delusion, before deciding this was straying too far 'off topic'. Eventually I concluded that, while physics was the *raison d'être* and generative force, it would become one of many things the work attended to, rather than its sole focus. In short, I succumbed to the magpie approach. The science is often submerged well beneath the surface: the line 'A smile is just fear with the corners turned up', for example, is a radical condensation of V.S. Ramachandran's theory about the evolutionary origins of the smile. Any scientific misconceptions or inaccuracies in the poem are entirely my own.

I don't suppose anyone expected that one of the writers involved with this project would actually attempt to tackle the question posed – quite casually, I suspect – in its title. At a certain point, however, I began to wonder how physics and neuroscience might account for such things as visions of angels and voices in the head. While reading up on the science, I also went back to a work that is a part of my literary DNA, the *Duino Elegies* of Rainer Maria Rilke (in which angels play a prominent role). The translation I have lived with longest, by US poet David Young, puts the Elegies into William Carlos Williams's triadic line or 'variable foot', and when I discovered that Williams was (at least in part) inspired by the theory of relativity to loosen up the poetic line in this way, the circle between science and poetry seemed to complete itself. I've used my own rough approximation for some sections of this work.

'Are Angels OK?' has a core cast of three men: Einstein, Rilke and Chaplin. It doesn't set out to present a complete or balanced portrayal of any of them; rather, it uses particular moments in their creative and personal lives for its own peculiar ends and argument.

Thanks

For the physics: Paul Callaghan, Pablo Etchegoin, John Hannah, John Lekner, Glenda Lewis, Howard Lukefahr, Jamie Sleigh, Alistair and Moira Steyn-Ross, Gillian Turner, Matt Visser, Richard Watts, Lara Wilcocks and Marcus Wilson. For the water, air and light: Sue and Barbara Allpress, Chris and Margaret Cochrane, Peter and Dianne Beatson, Dinah Hawken, Bill Manhire and Marion McLeod. For the opportunity: the International Institute of Modern Letters, the Royal Society and the Smash Palace Fund.

Acknowledgements

'Are Angels OK?' redeploys many of the motifs and images in Rainer Maria Rilke's *Duino Elegies,* with a particular leaning towards the Fifth Elegy. Where the Elegies are quoted from directly, the source is David Young's translation (Norton, 1978). The quotes from Einstein and his son Eduard are from *The Private Lives of Albert Einstein*, by Roger Highfield and Paul Carter (Faber & Faber, 1993). In Section I (ii), the first two lines are from Rilke's *The Notebooks of Malte Laurids Brigge* (1910); the other quotes are from his letters to his wife Clara as reproduced in *The Sacred Threshold: A Life of Rilke* by J.F. Hendry (Carcanet, 1983). The quote in I (iv) is from the Fifth Elegy. The description of Chaplin ('if you split him . . .') in II is from his friend Thomas Burke's essay 'A Comedian', quoted in David Robinson's *Chaplin: His Life and Art* (Collins, 1985), as is the phrase 'a hard, bright, icy creature' in IV. The lyrics in II are from the little-known introductory verses of Herman Hupfeld's classic, 'As Time Goes By'. My thanks to Moira Steyn-Ross for drawing these to my attention. The first Rilke quote ('The scale of the human heart . . .') in III is from a letter written during the First World War; the second is from the Sixth Elegy. The sentence about Max Brod in IV adapts a quote from his novel *Tycho Brahe's Path to God* found in the same volume. The Czeslaw Milosz quote in IV is from 'Craftsman' (*New & Collected Poems 1931–2001*, Ecco, 2001), and the quote in VII is from *A Treatise on Poetry* (Ecco, 2001), both translated by Milosz and Robert Hass. The quote in V is from the Second Elegy. The sentence referring to Robert Lowell in VII adapts lines from his poem 'The Literary Life, A Scrapbook', originally published in *Notebook* (1970) and revised in *History* (1973). The Hass quote is from 'Interrupted Meditation' (*Sun Under Wood*, Ecco, 1996). The Crackpot Index ('a simple method for rating potentially revolutionary contributions to physics') mentioned in VII can be found at http://www.phys.psu. edu/~scalise/misc/crackpot/crindex.html.

JO RANDERSON

Jo Randerson is one of the most gifted writers – and performers – in contemporary New Zealand theatre. She is the author of numerous plays including *The Unforgiven Harvest* and *Fold*, and is the founder and artistic director of Barbarian Productions. Her plays have been performed around New Zealand and overseas, and in 1997 she received the Sunday Star Times Bruce Mason Award for playwriting. Her collection of stories and fables, *The Spit Children*, was published in 2000. Her second book, *The Keys to Hell*, appeared in 2004. Jo was the 2001 Burns Fellow at Otago University, and was shortlisted for the 2006 Prize in Modern Letters. She has just gone to Europe to join a circus.

Jo Randerson writes:

'Everything We Know' is adapted from a theatrical lecture presented for the *Are Angels OK?* project. As an author of plays, short stories and stand-up comedy, it was with some degree of surprise that I found myself writing a piece of what I suppose may be described as non-fiction. It does not purport to be an essay, mainly because I was wary of trying to come to any definitive conclusion, although I was tempted.

Much was made during this project of the difference between art and science, with each of us keen to pin down some crucial definition. I see many links between not only the processes of the disciplines but also the areas of knowledge sought after – we are all looking for patterns, answers, stories. Why did Einstein talk of artists and scientists in the same breath and why in the past was there a much closer relationship between art and science, religion and even magic? Who is responsible for splitting these worlds?

Scientific observation reflects my own experience of humanity. Wealth is congregated in ugly fatty deposits. Poverty similarly distils in dark, uncomfortable crevasses. Nature seeks to balance this, as with osmosis, as with the second law of thermodynamics.

This project has been thoroughly enjoyable and challenging, and this piece of writing is by no means the end of the collaboration. Furthermore, I believe that communication and fearless dialogue is a very good start; and finally, because Bill asked in the first place, 'Are angels OK?', I would rather ask, 'Are the angels OK?' And my response is, 'I'm not sure, why don't we ask them?'

TONY SIGNAL

Tony Signal was born in 1962 and grew up in Feilding. After attending Massey University, he gained a PhD in theoretical particle physics at the University of Adelaide. In 1990 he was appointed to the lecturing staff at Massey University, and he became Professor of Physics in 2002. His main area of research is in understanding how matter is made from quarks. He is married and has two children.

DAVID WILTSHIRE

David Wiltshire is a Senior Lecturer in the Physics and Astronomy Department, University of Canterbury, Christchurch. He ensured the viability of *Dead of Night* in scientific terms, and contributed text for the sections involving *Endeavour*'s journey. He grew up in Palmerston North and studied theoretical physics at Canterbury and Cambridge universities. Following a PhD in Stephen Hawking's group in the 1980s, he held research and teaching appointments in Italy, the UK and Australia. He returned to New Zealand in 2001. His research interests are in black holes, quantum gravity and theoretical cosmology.